VEDANTA

Seeking Enlightenment Without a God

DR. LEO RASTOGI

Anahata Books
The Woodlands, Texas

Printed in the United States of America
Published in 2024 by Anahata Books
1st paperback edition, 2024

1st Edition

Paperback ISBN: 978-1-7358714-4-8

www.leorastogi.com

This book is dedicated to the lineage of Vedic Rishis, teachers, and students who have kept the source of eternal wisdom of the Vedas alive.

ACKNOWLEDGMENTS

This book is, first and foremost, a humble effort on my part as a lifelong student of the Vedas to pass along my own personal synthesis.

In this endeavor, I have been helped by many teachers at whose feet I have sat, and by countless conversations with fellow learners whose insights have benefited me tremendously. Additionally, my students, whose curiosity has inspired my continued learning, have also played a significant role. While it is simply not possible for me to name all of them, I would like to begin by thanking them all. But for their contribution, this book would not be possible.

Next, I want to thank my family, my spiritual coworkers and friends, who have supported this endeavor in their own way. Thank you Dr. A.P. Rastogi, Dr. Rekha Rastogi, Nirjhara, Richa, Javi, Charlie, Rodrigo, Ale, Alice, Karol, Suzanne, I am truly grateful.

For a book to be a finished product, it needs the support of a great editing and production team, and I was blessed to have the best. Thank you Wendi, Alina, and Dmitry for making this look so easy!

Finally, and most importantly, I want to thank my esoteric mentors and invisible guides, who have been a constant presence, guiding each step of my own journey. Truly humbled and grateful for your support, I am.

CONTENTS

CONTENTS

"Most of my ideas and theories were heavily influenced by Vedanta."

– Erwin Schrödinger

PREFACE

Welcome to the world of Vedanta — a realm of ancient wisdom that beckons us to explore the depths of our existence and seek enlightenment beyond the boundaries of traditional beliefs. Picture yourself standing at the edge of an expansive ocean, where the waves of time brush against the shores of human history, in fact even cosmic history. In this vast expanse, we are at the threshold of eternity, poised to embark on a journey of self-discovery and transformation.

Let me begin by sharing a personal anecdote. I first discovered Vedanta at the age of twelve, as I was navigating questions about *'who created everything'* and looking for answers in science, only to find my way to this very rich reservoir of wisdom. This, however, happened most unexpectedly. I was reading about one of my heroes, the great physicist, and Nobel laureate

Erwin Schrodinger (of Schrodinger's cat fame). In his book *What is Life?* he writes, "*The unity and continuity of Vedanta are reflected in the unity and continuity of wave mechanics. This is entirely consistent with the Vedanta concept of All in One.*" I guess I was especially intrigued because at that time, I was struggling to follow the mathematics at the source of his famous equation and felt that maybe studying Vedanta would help! A piece of trivia: while Schrodinger's cat is famous for being dead and alive at the same time (an example of superposition), one fact very less known is that he had named his dog *Atman* and used to call *Atman=Brahman,* as Schrodinger's second equation.

Interestingly, this was not the only time I ran into a physicist pointing to the Vedas. I also came across Heisenberg, the great physicist and Nobel laureate, having said, "*After the conversations about Indian philosophy, some of the ideas of Quantum Physics that had seemed so crazy suddenly made much more sense.*" *Most significantly, I learned that Niels Bohr, another Nobel laureate, stated, 'I go to the Upanishads to ask questions.'*"

This also makes me wonder sometimes, when the schism between Vedanta and Physics really occurred. So much so that if someone talks about consciousness, they are guaranteed never to get tenure in a Physics department!

Having had the privilege of being a lifelong student of Vedanta, I am continually amazed by one profound realization. My personal quest has encompassed a wide range of subjects, from quantum physics and abstract mathematics to cognitive psychology and existential philosophy. Remarkably, many of these disciplines either stem from or converge upon the principles of Vedanta. It is, therefore, my humble privilege to accompany you on your journey through this book.

Let me now move on to discuss the thinking behind the title. After all, God has been the central character in the grand theater of humanity's history, so deeply entrenched that imagining life without such a presence seems unfathomable to many. This sentiment rings especially true for those steeped in Western religious traditions, where belief in a singular, omnipotent God forms the bedrock of faith.

But let's rewind the hands of time to an era before the stage was set with organized religions. Picture a world where our ancient ancestors, far removed from the doctrines of gods and deities, roamed the earth. From the emergence of Homo Sapiens over 300,000 years ago to the rise of sophisticated civilizations, divine powers hardly held sway.

In those primal days, human tribes, bound by the rhythms of nature, lived intimately connected to the world around them. They wove spiritual significance into every rustle of the wind, every creature's call, every celestial dance. Animistic beliefs flourished, casting a mystical aura over natural phenomena, animals, plants, and the stars above. In the face of nature's might, they stood humbled and awestruck, recognizing a force beyond their comprehension.

Fast forward to the dawn of settled societies, around 10,000 years ago. As our ancestors traded nomadic life for settled agrarian communities, the seeds of complex religious beliefs began to sprout. The domestication of plants and animals heralded the birth of farming, leading to permanent settlements and the dawn of civilizations. Amidst this burgeoning complexity, organized religion took root, transforming nature's raw power into an intricate quilt of polytheistic worship.

With each passing era, humans began to anthropomorphize the forces of nature, assigning them human-like qualities and identities. They crafted a pantheon of gods and goddesses, each personifying different facets of existence. This marked the genesis of what we might call the "me+god" era, where humanity placed its ego right alongside the cosmic order. Rituals, ceremonies, and communal gatherings became the canvas upon which they painted their beliefs, fostering social unity and cultural identity.

Yet, the journey didn't end there. Millennia passed, civilizations rose and fell, and belief systems evolved. Monotheism emerged as a latecomer to the spiritual stage, with major world religions like Judaism, Christianity, and Islam championing the supremacy of a single, all-encompassing God. But before the advent of monotheism, ancient civilizations like the Sumerians, Babylonians, Egyptians, and Greeks worshiped a dazzling array of gods and goddesses, each crafting their own celestial narratives.

Amidst this kaleidoscope of spiritual evolution, a different tale unfolded in the Indus Valley region. Here, the early Vedic system of belief blossomed, marking what scholars term the Vedic period. Assimilating within it indigenous beliefs and those brought in by migrants from Central Asia, ancient India became a melting pot of beliefs, rituals, and cosmic reverence.

The sacred texts of the Vedas served as vessels for this spiritual wisdom, preserving hymns, prayers, and rituals dedicated to a myriad of deities. Indra, Agni, Varuna, Surya — these were not mere names, but embodiments of divine forces invoked for blessings ranging from victory in battle to bountiful harvests.

In the cradle of Vedic thought, sages delved deep into the mysteries of existence, probing the nature of reality, self, and the ultimate purpose of human life. The Upanishads, revered for their profound insights, offered seekers a roadmap to spiritual enlightenment — moksha — the liberation from the cycle of birth and death. Here, enlightenment wasn't just a distant ideal but a tangible goal, beckoning souls to transcend the confines of the ego and embrace their eternal essence.

Unlike the Mesopotamians, whose afterlife was steeped in ritual and divine favor, Eastern faiths offered a different path — one of self-realization and inner illumination. This divergence laid the foundation for what centuries later became the East-West schism in spiritual perspectives, each forging its own narrative of salvation and enlightenment.

But amidst the diversity of religious traditions, a common thread binds humanity's quest for enlightenment. Whether through theistic devotion or introspective inquiry, the pursuit of truth and self-realization transcends cultural boundaries. And in this tapestry of spiritual exploration, Vedanta emerges as a beacon of universal wisdom — a call to embark on a journey of inner discovery, unshackled by dogma or doctrine.

So, dear reader, as you turn the pages of this book, I invite you to embark on a quest beyond the confines of traditional religious paradigms. Peel back the layers of your inner landscape, confront life's existential mysteries, and embrace the transformative wisdom of Vedanta. Let this be the beginning of a profound journey — one that leads you not only to self-discovery but to unity with the cosmos itself.

Welcome to the adventure of a lifetime!

"Each soul is potentially divine. The goal is to manifest this divinity within by controlling nature, external and internal. Do this either by work, or worship, or psychic control, or philosophy — by one, or more, or all of these — and be free. This is the whole of religion. Doctrines, or dogmas, or rituals, or books, or temples, or forms, are but secondary details."

– Swami Vivekananda

"Vedanta is the teaching of non-duality, which means that all is one and that the Divine is present in all things."

– David Frawley

CHAPTER 1

BEYOND GODS AND GODDESSES: THE NON-THEISTIC ESSENCE OF VEDANTA

Introduction

As we commence our conversation about Vedanta and its "non-theistic" essence in earnest, it's a good time for us to briefly consider what God means, at least in the context of the chapter title.

God is perhaps one of the most complex concepts that humanity has come to birth, perceive, experience, and build a relationship with over millennia. A deep discussion about the subject is beyond the scope of this book and merits an entire work of itself. In fact, I have personally written such a book titled *"Many Paths, Many Truths: A Journey through world religions that have shaped 21st century humanity"* with the intent of providing exactly such an immersion, and I obviously highly recommend it.

For the purpose of this chapter, let me just say that the very meaning of this word "God", which most of us find ourselves using often in our lives, has evolved over thousands of years. It may represent a supernatural power or an all-powerful creator; a pantheon of millions of deities or just "a human idea", depending on your own personal beliefs. As varied as the descriptions of God might be, there seems to be a common theme across the history of humanity: that this being called God has a major say in *our reward for (or punishment for not) living a virtuous life, salvation (or some equivalent) at the end of it, the nature of what lies before birth and after death, and how the universe came to be*. If simply said, beliefs associated with who God is and what he/she wants us to do with our lives are supposed to point us to the north of "living our best life". Most world religions or systems of faith have such objects of devotion in either a monotheistic, polytheistic, animistic, or naturalistic framework.

This is where the teachings of the Vedas and Vedanta, therefore, are very unique. They present to us a very wholesome philosophy of life, the birth of the universe, the nature of consciousness, etc. All this, within a framework almost wholly independent of a "conventional God", as we have come to understand it in modern religions, including the post-puranic polytheistic version of Hinduism.

This also directly implies that the nature of "enlightenment" as the end goal of a spiritual awakening in Vedanta is very different from what is understood by the meaning of this word in the context of Western systems of belief, philosophy, or the English language.

Much of that, of course, has to do with my favorite complaint – that the English language does not provide a rich enough vocabulary to appropriately translate from profound spiritual Sanskrit or philosophical concepts. In original Vedic and Upanishadic texts, we have words like *Mokha, Nirvana, Mukti, Kevalya,* etc., that speak to distinct experiences of spiritual awakening or transcendent realizations – none of which translate well into English ☺.

Therefore, I propose that we begin this chapter by developing a foundational understanding of what enlightenment means in Vedanta, and that is what we will explore in the next few pages.

Before we get to that, I also wish to clarify a very important concept. When in the book I refer to "Hinduism",

I am not necessarily referring to the current prevalent temple-based Hindu practices that we see widely popular in India. Instead, I am actually referring to the original system of spiritual and metaphysical beliefs, teachings, and philosophy as laid out in the Vedas and Upanishads (more on that later) in ancient India (think 6000 BCE or older).

While some sects within Hinduism still follow the puritanical system of Vedic practices, the religion itself (which it has now certainly become) has gone through many cycles of evolution over the last 2000 or so years, with many new beliefs and practices added. Much of this happened as the original Vedic religion had to deal with several seemingly threatening developments, such as the rise of Buddhism, Jainism, the influence of Islam, the arrival of Christianity, etc. This journey can also be explored in detail in my other book I mentioned earlier. I will also provide some additional reading references for the most committed or curious readers. But I must quickly reiterate that this book is not about Hinduism; this book is about Vedanta – one of the many threads of the beautiful mosaic of Hinduism!

Having said that, I have used Hinduism as a catch-all term in this book for the ease of the reader to represent Vedic beliefs, Upanishadic beliefs, and Sanatana Dharma*. Nonetheless, I have tried my best to make as much distinction as I can with references to the texts.

Demystifying Enlightenment: *Age of Reason* in the West vs. *Spiritual Awakening* in the East

Let me begin by posing a question: *What does "enlightenment" mean to you?*

There are so many ways to approach this conversation, and it can mean many different things to different people. *Is it aligning with your inner self, connecting with the universe, an existential transformation, something more mystical and transcendental, or simply the pursuit of knowledge?*

Whatever your take on the subject, it is probably infused with some spiritual yearning or experience. If you are on a quest for enlightenment – which is deeply personal, you may or may not have included God or gods in that equation. The idea of enlightenment without a deity may seem strange to you, but it is one that has been distilled for thousands of years in the East, where the philosophy of Vedanta was born.

The term "enlightenment", as we said earlier can mean many different things, even trigger confusion, especially if you come from the West, where it hasn't always been associated with spiritual awakening. Indeed, for several centuries, it was used as another word for "reason", which is rather strange to the Eastern mind. For a Hindu or follower of Vedanta, enlightenment means something completely different and couldn't be further removed from concepts of logic or reason associated with European thought.

If we look at the European experience of Enlightenment, it goes back to the intellectual and cultural movement of the 17th and 18th centuries; the Age of Enlightenment (also known as the Age of Reason). It did not have religious or spiritual connotations in the way we may think of the word "enlightenment" now but referred to intellectual and philosophical movements that emphasized reason, empirical inquiry, and the advancement of knowledge.

Enlightenment figures such as René Descartes, John Locke, Voltaire, Immanuel Kant, Jean-Jacques Rousseau, and Adam Smith shaped intellectual history, promoting critical thinking and progress, though the movement also affected spirituality. Although some thinkers of the Enlightenment period did engage with religious ideas and critique religious authority, the concept of enlightenment itself was more closely tied to secular notions of progress, rationality, and human autonomy. It was about shedding superstition and dogma in favor of critical thinking and scientific understanding.

It's as if Europeans turned their backs on any notions of spiritual awakening and, instead, embraced the shiny new ideas filtering into the public psyche through the works of its influential thinkers. No stone was left unturned in the pursuit of reason across the fields of philosophy, science, politics, literature, and economics. This emphasis on reason and empiricism led to increased skepticism toward religious beliefs and spiritual experiences that couldn't be explained or verified through empirical means, leading to a decline in religious influence over political, social, and cultural institutions. As a result, spirituality became increasingly marginalized in public life, with mystery, transcendence, and awe getting lost.

It wasn't until the 19th and 20th centuries that the concept of spiritual enlightenment began to filter into Western thought as a result of increased global interconnectedness, cross-cultural exchange, and the study of Eastern philosophies and religions. Several key protagonists played significant roles in introducing and popularizing ideas related to spiritual enlightenment in the West, such as Ralph Waldo Emerson, Henry David Thoreau, Helena Blavatsky, Swami Vivekananda, D.T. Suzuki, and Alan Watts.

Spiritual Enlightenment and Vedanta

Many philosophical and spiritual traditions, such as Buddhism, certain forms of Hinduism, Taoism, and secular humanism, offer pathways to enlightenment that do not rely on a personal deity. Rather, these paths emphasize inner exploration, meditation, mindfulness, ethical living, and the cultivation of wisdom and compassion. God or gods are taken out of the equation.

Seeking enlightenment without a belief in God involves exploring the depths of one's own consciousness, understanding the nature of reality, and finding meaning and purpose in life without reference to a divine being. It's a deeply personal journey that you can pursue through various practices and philosophies, depending on your beliefs and inclinations.

Such a journey, however, is much accelerated if we have the ability to engage with truly profound ancient wisdom. One of the most significant sources of this nature is the Vedas, as evolved in ancient India. We will discuss them at quite some length later in the book but for the moment, suffice to say that, as ancient as they are (some estimates put them at between 1500-6000 BCE), their wisdom seems timeless. They have inspired hundreds of generations of humanity, including our modern-day thinkers, philosophers, and scientists alike.

Written in Sanskrit, the Vedas are mostly outside the reach of a normal person. In addition to that, they are very esoteric in nature, needing a guide to truly make sense of them as a genuine philosophical and spiritual immersion. Vedanta is a system of Philosophy, or shall I say, a collection or ocean of systems of Philosophy, that bridges the gap between the esoteric revelations, rites, and rituals of Vedas and the quest for meaning that humanity shares.

In this book, I want to introduce you to Vedanta not just as a philosophy but also as a potential path to spiritual enlightenment. It would be a mistake not to point out that Vedanta on its own is one of the six schools (some argue seven schools) of Vedic philosophy and is one of the most influential philosophical traditions in India. Each of the other six *Astik (that believe in the supremacy of Vedic knowledge)* philosophical traditions are also equally rich, but that is a whole other book!

A note about the history of the Vedas

Up until fairly recently, it was believed that a group of Indo-European-speaking nomads known as the Aryans migrated into the Indian subcontinent around 1500

BCE. They settled in the Indus Valley region, bringing with them ancient wisdom that was eventually laid out in the Vedas. Often referred to as the Aryan Migration or Invasion Theory, this idea was developed in the 19th century by Western scholars but experts now know that this theory is certainly flawed.

The Aryan Migration Theory was largely formulated based on linguistic studies, particularly the similarities between Sanskrit and other Indo-European languages such as Latin, Greek, and Persian. Scholars like Max Müller expanded on these linguistic connections and proposed that the Aryans were a distinct group who migrated into India and brought with them the Vedic culture. Early archaeological findings were interpreted to show a discontinuity between the Indus Valley Civilization (Harappan culture) and the later Vedic culture, suggesting a new group of people (the Aryans) had arrived and influenced the region.

In turn, the Aryans were traditionally considered the founders of Vedic thought, bringing with them the Sanskrit language and Vedic culture, which significantly influenced the Indian subcontinent's social and cultural landscape.

Now, these theories are being questioned as research into genetics indicates either a migration from Central Asia or a more complex demographic history involving multiple waves of migration and interaction. The archaeological evidence is also open to speculation, with excavations and findings from the Indus Valley Civilization having been interpreted in various ways. Some scholars see continuity between the Indus Valley Civilization and later Vedic culture, while others point to differences in material culture and urban planning as evidence of a new cultural influx.

In political and cultural terms, the debate also has significant ramifications. The indigenous Aryan theory is often linked with nationalist narratives that emphasize the ancient and continuous development of Indian civilization, while the migration theory is sometimes viewed as diminishing the antiquity of Indian culture.

My own research on this subject, including the study of the ancient spiritual sites and mapping continuity of spiritual practices around the world, combined with developing intuitive insights from Esoteric lineages, has brought me to the working conclusion that the so-called Aryans were almost certainly indigenous to the subcontinent. There is clearly a continuous thread of spiritual systems that traces back thousands of years, of which the Vedas are but a temporal compilation. But ultimately, this debate remains unresolved, with strong arguments on both sides and ongoing research continuing to shed new light on the complex history of the Indo-Aryans.

Vedanta: The synthesis of Vedic wisdom

The term "Vedanta" is derived from two Sanskrit words: "Veda," which means "knowledge" or "sacred scripture," and "anta," which means "end" or "culmination." Thus, Vedanta can be understood as the culmination or end of Vedic wisdom or, as I like to say, the synthesis of Vedic wisdom.

That wisdom obviously refers to the Vedas, which consist of four main collections: the Rig Veda, Sama

Veda, Yajur Veda, and Atharva Veda (more on that later), along with associated texts known as Brahmanas and Aranyakas. The Vedas contain hymns, rituals, prayers, and philosophical speculations dating back to ancient India. The core focus of the Vedas is mostly on rites and rituals, chants and prayers, customs, and applications, with very little philosophical or explanatory content – which is the focus of the Upanishads.

As I shared earlier, the nature of the Vedas is very esoteric and difficult to understand even for experts. This eventually led to the emergence of dogmatic practices of following Vedic rituals without understanding their meaning or significance. Over time a new body of scriptures/knowledge emerged in ancient India called the *Upanishads.* The purpose of these texts was to extrapolate, explain, and bridge the Vedic texts and knowledge into a more commonly explorable system of philosophy related to Spiritual Enlightenment.

The Upanishads delve into metaphysical questions about the nature of reality, the self, and the ultimate truth (Brahman). They offer a vision of an interconnected universe with a single unifying principle behind the apparent diversity in the cosmos, presenting various philosophical insights and spiritual teachings. They are truly the last section of revelation of the Vedas. In ancient India, the Upanishads were taught to a student of the Vedas at the end, perhaps because they deal with the ultimate goals of the Vedas – *achieving Moksha, supreme bliss, or supreme enlightenment.*

Flowing from the growth of the awareness of the Upanishads, Vedanta emerged as a distinct spiritual philosophical tradition in India as one of the six Astika philosophical traditions. *(Astika loosely means a tradition that believes in the supremacy of the Vedas).* Vedanta focuses on the diligent and extensive interpretation of the philosophical teachings found in the Upanishads. It has, over time, also developed into several distinct schools with somewhat differing interpretations of the Upanishads, which we will briefly discuss later in the book.

Vedanta is, therefore, somewhat of a culmination, endpoint, or a synthesis (as I like to call it) of Vedic wisdom for the following reasons:

Philosophical Sophistication:
While the Vedas contain a rich array of rituals, hymns, and practical teachings, Vedanta represents a more sophisticated and abstract philosophical inquiry into the nature of reality. The Upanishads, as the philosophical portions of the Vedas, delve deeply into metaphysical questions and present profound insights into the nature of existence.

Synthesis and Interpretation:
Vedanta synthesizes and interprets the teachings of the Vedas, particularly the Upanishads, offering a comprehensive philosophical framework that addresses fundamental questions about the nature of reality, the self, and the ultimate truth. It builds upon the foundational teachings of the Vedas and provides a deeper understanding and interpretation of their wisdom.

Spiritual Culmination:
Vedanta presents the highest goal of human life as spiritual realization (moksha) and offers teachings and practices aimed at achieving liberation from the cycle

of birth and death. In this sense, Vedanta represents the ultimate goal or endpoint of the spiritual journey outlined in the Vedas.

At its core, Vedanta seeks to interpret and synthesize the philosophical teachings found in the Upanishads, which are considered the concluding portions of the Vedas, the oldest scriptures of Hinduism.

The key concepts within Vedanta include:

- **Brahman:** The ultimate reality or divine essence underlying all existence. Brahman is described as eternal, infinite, and beyond all attributes and distinctions. It is considered the unchanging, absolute reality from which the universe emanates.
- **Atman:** The individual self or soul. Vedanta teaches that the true nature of the individual self (Atman) is identical to Brahman, the ultimate reality. Realizing this identity, known as "atma-jnana" or self-knowledge, is considered the goal of human life.
- **Maya:** The principle of illusion or cosmic ignorance that veils the true nature of reality. Maya is seen as the cause of the mistaken perception of diversity and separation in the world. Spiritual liberation (moksha) involves transcending Maya and realizing one's unity with Brahman.
- **Jnana Yoga:** The path of knowledge or wisdom. Vedanta emphasizes the importance of self-inquiry, philosophical contemplation, and the study of scripture as a means to attain spiritual awakening and liberation.

- **Bhakti Yoga and Karma Yoga:** While Vedanta primarily focuses on knowledge and understanding (jnana), it also acknowledges the importance of devotion (bhakti) and selfless action (karma) as paths to spiritual growth and awareness.

Vedanta has various philosophical sub-schools or interpretations, including Advaita Vedanta (non-dualistic), Dvaita Vedanta (dualistic), and Vishishtadvaita Vedanta (qualified non-dualistic), each offering different perspectives on the nature of reality and the relationship between the individual soul and the ultimate reality. This is not an exhaustive list – there are several other sub-schools.

However, the non-theistic essence of Vedanta refers to the core philosophical principle that underlies many interpretations within the Vedanta tradition, particularly Advaita Vedanta. Founded by the philosopher Adi Shankaracharya, Advaita Vedanta emphasizes the non-dualistic nature of reality and the ultimate identity between the individual self (Atman) and the ultimate reality (Brahman).

In Advaita Vedanta, Brahman is considered the only reality, transcending all dualities, including the distinction between creator and creation, subject and object, or God and individual souls. While Brahman is sometimes described using theological language or metaphorical terms in religious contexts, Advaita Vedanta ultimately teaches that Brahman is beyond all attributes, forms, and conceptualizations.

From a non-theistic perspective, Vedanta emphasizes that Brahman is not a personal deity in the conventional sense, nor is it limited to any particular religious conception of God. Instead, Brahman is the impersonal, formless, and all-pervading reality that encompasses everything in existence. In this view, the ultimate goal of spiritual practice is not necessarily to worship or seek the favor of a deity but to grasp one's essential unity with Brahman through self-knowledge (atma-jnana) and direct experience (anubhava).

While Vedanta acknowledges the existence of deities and various forms of religious worship and devotion, it teaches that these are ultimately means to an end — tools for spiritual growth rather than ends in themselves. The non-theistic essence of Vedanta invites practitioners to transcend anthropomorphic conceptions of divinity and recognize the underlying unity of all existence.

It's important to note that while Advaita Vedanta presents a non-theistic understanding of ultimate reality, other interpretations within the Vedanta tradition, such as Vishishtadvaita Vedanta and Dvaita Vedanta, incorporate theistic elements and posit a personal God who is distinct from the individual soul. However, even within these interpretations, the ultimate goal remains spiritual awareness and the recognition of the essential unity of all existence.

I want to mention here that although Advaita Vedanta may emphasize the non-dualistic understanding of Brahman and Atman, it does not negate the validity of personal devotion or the worship of deities. Vedanta recognizes that individuals have different temperaments and spiritual inclinations and therefore offers various paths (yogas) to suit the diverse needs of seekers. Whether one approaches the divine as an impersonal principle (Brahman) or as a personal deity (Ishta Devata), the ultimate goal remains the same.

You may find it difficult to get your head around the notion of Vedanta if your background lies within the Western tradition of God. It could be that you have been influenced by Judeo-Christian traditions and deeply ingrained beliefs in a personal God who possesses human-like attributes, such as intelligence, will, emotions, and the ability to intervene in human affairs. This anthropomorphic conception of God can be found in Western religious teachings, literature, and art, making it challenging to conceive of a divine reality that transcends human form and attributes.

I think it's also fair to say that Western languages and concepts may lack adequate vocabulary and frameworks to articulate or understand the non-anthropomorphic aspects of divinity. Terms like "God" or "the Divine" often carry anthropomorphic connotations in Western discourse, making it difficult to grasp the idea of an impersonal, formless, and attribute-less ultimate reality.

In addition, we have to consider cognitive biases because, as human beings, we tend to anthropomorphize phenomena and attribute human-like qualities to abstract concepts, including the divine. This urge to do so is rooted in cognitive biases and social conditioning, making it challenging to conceive of a divine reality that transcends human attributes and limitations. Challenging these deeply ingrained beliefs and exploring alternative conceptions of divinity can be testing for many people.

And, we shouldn't forget that Western societies have a habit of prioritizing individualism, rationalism, and empirical inquiry, which may lead some individuals to approach religious and spiritual matters from a more analytical and skeptical perspective. The notion of an impersonal God may seem abstract or esoteric within this cultural context, and it requires a shift in mindset and worldview to fully grasp the concept.

However, with an open mind, sincere inquiry, and exposure to diverse philosophical perspectives, we can expand our understanding of spirituality and appreciate the richness of the conception of divinity as presented in Vedanta.

Let's begin that inquiry!

As a first step, let's take a closer look at the Vedas to gain a greater understanding of these ancient texts.

The schooling I received from a nine-year-old

I was in my late twenties when I met one of my spiritual mentors in India, who has been a great influence on my life. When I had studied with him for a while, as the tradition suggests, I asked him what I could offer him for this *Gurudakhshina (tuition).* He asked me to support him in increasing the reach of his message to the "new digital world", which, essentially, meant making his lectures, meditations and books available in digital format.

To make progress on this, I contacted some of my friends in the media industry in Mumbai, which I had understood was the center of creative arts. I made several trips there with my teacher and, on one of those trips, we were invited for lunch by a friend of mine.

When we arrived at his place, we understood that my friend's family was expecting to meet a Guru – an enlightened master – and had an obviously deferential behavior towards my mentor.

However, it soon became clear to them that he was not a typical "guru God" that Indians are used to but *a teacher* in the best sense of the word. Once this slight disappointment was over, we had a very nice chat after lunch about the role of spirituality in our lives, including conversations about the nature of the soul, etc.

Playing in the living room, where we were seated at this point, was also my friend's nine year old daughter. After a while, she began to listen to our conversations intently while we were enjoying a second serving of ice-cream.

My teacher then decided he wanted to rest a bit and was taken to the guest room, while I continued to chat with my friend and his family. Suddenly, out of nowhere, my friend's daughter asked me a question, setting off a wonderful inquiry. I am reproducing below our exchange, as accurately as my memory can fathom after almost two decades.

"Uncle Leo, can I ask you a question?"

"Of course!"

"Have you seen God?"

"It depends on who you call God."

"I mean, like you have seen me and Papa, the statue of lord Ram in the temple. Like that, with a face and body and all as I see in my books."

"No, I have not seen God like this and, also, I do not believe in a personified God but in a higher presence that pervades everything in the universe."

"So, you have not seen him and have no idea what he looks like. How do you know he is there?"

"Well, actually, I don't think it's a person. But I have felt his presence."

"How ?"

"In my meditation I get insights, deep realizations, I even feel at one with that presence at times."

"Hmm, that is interesting. You know what I think, Uncle?"

"No, please tell me."

"You have not seen him, You don't know what he looks like, but you get insights from him. So I think you are your own 'God. That's why you feel at one with him, because he is inside you."

"Wow... maybe you are more right than you think, kiddo! Thanks for reminding me."

"The Vedas are the most precious gift for which the West has ever been indebted to the East."

– Max Müller

CHAPTER 2

THE VEDAS: SACRED WISDOM OR DIVINE REVELATIONS

The word Veda comes from the ancient Sanskrit word "Vid", which means "to know". The Vedas are ancient texts first emerging as a compilation around 1500 BCE and later codified around 300 BCE. Their origin, however, is almost certainly thousands of years earlier, with some conservative estimates putting their composition dates at older than 6000 BCE.

The Vedas are revered as divine revelations and believed to have been directly heard (Shruti) by ancient sages or seers known as ***Rishis***. They are written in Sanskrit and are arguably the oldest texts in any Indo-European language. The Vedas are often considered to be the foundational text of Sanatana Dharma*, which, in its modern form, could be said to be Hinduism. To that extent, one could say that the Vedas are the foundational scriptures of Hinduism.

While it's true that they are written in Sanskrit, it is important to know that, unlike classic Sanskrit, the Vedas are actually written in archaic Vedic Sanskrit. This is a language difficult to understand even for modern Sanskrit scholars, as it predates the formalization of Sanskrit grammar by Panini.

Most of the original translation of the Vedas was done by German, English, French, and other scholars without a dictionary, so the accuracy of early translations into English is debatable. There are, however, later translations and interpretations by contemporary Indian scholars which I have personally found qualitatively different.

Having been armed with a decade of Sanskrit learning in school, I have painstakingly labored through sections

*Chart Depicting Components of Vedic Literature

Vedic Literature
- Śruti
- Smṛti

Śruti

4 Vedas
- Rg
  - Saṃhitā
  - Brāhmaṇa
  - Āraṇyaka
  - Upaniṣad
- Sāma
- Yajur
- Atharva

Each Veda has its separate Saṃhitā, Brāhmaṇas, Āraṇyakas, Upaniṣads

Smṛti

6 Vedāṅga
- Kalpa
  - Śrauta Sūtras
  - Gṛha Sūtras
  - Dharma Sūtras
- Śikṣā
- Chanda
- Nirukta
- Vyākaraṇa
- Jyotiṣa

4 Upavedas
- Ayurveda
- Arthaśāstra
- Dhanurveda
- Gandharvaveda

6 Darśanas
- Nyāya
- Vaiśeṣika
- Sāṃkhya
- Yoga
- Pūrva-Mīmāṁsā
- Uttara-Mīmāṁsā

Purāṇas	18 Purāṇa and 18 Upapurāṇa
Itihāsa	Rāmāyaṇa and Mahābhārata
Dharmaśāstra	Several texts

*Adapted from : Hinduism Ritual, Reason and Beyond by Ashok Mishra (2023)

of the Vedas and Upanishads with the full benefit of a dictionary and the help of Google. Despite that, I still find the esoteric meaning and implications of the Vedas require contextual interpretation and imagination as much as they require linguistic skills. Furthermore, at different points in time, I have revisited the same sections of the Vedas (and Upanishads) and come away with a completely new and profound understanding than previous journeys – each time feeling more enriched.

The 4 Vedas

The Vedas are traditionally divided into four main collections, each with its own specific focus and content:

1. **Rig Veda:** The oldest and most revered of the Vedas, consisting of hymns and prayers addressed to various deities such as Agni (fire), Indra (king of gods), Varuna (god of cosmic order), and others. The Rig Veda is divided into ten books, known as Mandalas, and contains a total of 1,028 hymns.

2. **Sama Veda:** A collection of melodies (saman) or chants derived from the hymns of the Rig Veda. The Sama Veda is primarily used by priests (udgatrs) during rituals and sacrifices. It is also divided into two main parts: the "Udgritha Samhita" and the "Upanishad Samhita."

3. **Yajur Veda:** A collection of prose mantras and sacrificial formulas used by priests (hotrs) during rituals and ceremonies. The Yajur Veda Is divided into two main branches: the "Shukla Yajur Veda" (White Yajur Veda) and the "Krishna

Yajur Veda" (Black Yajur Veda), each with its own distinctive style and content.

4. **Atharva Veda**: A collection of hymns, incantations, and spells used for various purposes, including healing, protection, and magical rituals. The Atharva Veda is considered somewhat distinct from the other Vedas due to its focus on practical and worldly concerns, as opposed to the more ritualistic and cosmological themes found in the other three Vedas.

Interestingly, in original writings, we only find references to three Vedas, with the fourth Atharva Veda appearing mostly around the time of Manu (200 BCE-100 CE). Manusmriti, the work of Manu, mentions only three Vedas, as do the early Buddhist works (~400 BCE). However, let this not take away from the immense importance and wisdom of the Atharva Veda, which has knowledge ranging from healing practices to cosmology!

Structurally, each Veda is further divided into four main parts:

1. **Samhitas:** The core collection of hymns, prayers, and mantras associated with each Veda.
2. **Brahmanas:** Prose texts that provide explanations and interpretations of the rituals and sacrifices described in the Samhitas.
3. **Aranyakas:** Texts that contain philosophical reflections, meditative practices, and symbolic interpretations of the rituals, often intended for forest-dwelling hermits (vanaprasthas).

4. **Upanishads:** Philosophical treatises that explore profound metaphysical and spiritual concepts, including the nature of reality, the self, and the ultimate truth (Brahman).

The Vedas can also be further divided into three main sections – the Jnana Kanda, Karma Kanda, and Upasana Kanda, which reflect their diverse teachings and purposes:

- **Jnana Kanda:** Also known as the ***"Knowledge Section",*** Jnana Kanda deals with the philosophical and metaphysical aspects of reality, exploring concepts such as the nature of existence, the self (atman), Brahman (the ultimate reality), and the paths to spiritual enlightenment (moksha). These texts delve into profound inquiries about the nature of the universe and the self, aiming to lead the seeker to a higher understanding beyond the material world.
- **Karma Kanda:** The ***"Ritualistic Section"*** of the Vedas, Karma Kanda primarily focuses on rituals, ceremonies, sacrifices (yajnas), and ethical duties (dharma). It provides guidelines for leading a righteous and ethical life, emphasizing the importance of performing one's duties without attachment to the results. Rituals prescribed in this section are believed to bring worldly benefits and spiritual progress when performed with the right intention and adherence to Vedic injunctions.
- **Upasana Kanda:** Also known as the ***"Worship Section"*** or ***"Bhakti Kanda,"*** Upasana Kanda

comprises hymns, prayers, and devotional practices aimed at fostering a deep connection with the divine. It emphasizes the worship of various deities (devas) through meditation, chanting, and other forms of devotional practices. This section emphasizes the importance of cultivating a personal relationship with the divine and experiencing spiritual fulfillment through devotion and surrender.

In Hindu tradition, the Vedas are categorized into two main types of scripture:

- **Shruti:** Literally means *"that which is heard."* Shruti scriptures are considered to be divinely revealed and eternal truths heard by ancient sages. The Vedas, including the Samhitas, Brahmanas, Aranyakas, and Upanishads, are collectively classified as Shruti. As heard here would probably be better translated as *"intuitively realized or as heard by inner listening/ear."*
- **Smriti:** Literally means *"that which is remembered."* Smriti scriptures are considered human compositions based on memory and tradition. They include texts such as the Itihasas (epics) like the Mahabharata and the Ramayana, the Puranas (mythological narratives), the Dharmashastras (legal and ethical texts), and various other treatises on philosophy, arts, and culture.

Let us now briefly immerse ourselves in some curated extracts from each of the Vedas.

1. Rig Veda

The Rig Veda is the oldest and most revered of the four Vedas, a collection of ancient sacred texts that form the foundation of Hinduism. Believed to have been compiled between 1500 and 1200 BCE, though some traditions suggest the hymns were composed much earlier, perhaps as far back as 6000 BCE, the Rig Veda offers profound insights into the early spiritual and cultural life of the Indo-Aryans in ancient India.

It is made up of 1,028 hymns, known as *suktas*, which are divided into ten books called *mandalas*. These hymns are dedicated to various deities and natural forces, reflecting the polytheistic and animistic beliefs of the time. The primary deities addressed in the Rig Veda include *Agni* (the fire god), *Indra* (the warrior god and king of the gods), *Varuna* (the cosmic order keeper), and *Surya* (the sun god), among others. These deities personify natural elements and cosmic principles. The hymns encompass a wide range of themes, including cosmology, mythology, rituals, prayers, and philosophical inquiries. They express gratitude, seek blessings, and offer invocations for prosperity, health, and protection.

The Rig Veda is central to Vedic rituals and ceremonies. Its hymns are recited during various religious rites and are considered to hold immense spiritual power. The text also provides the liturgical foundation for other Vedic scriptures. It offers a window into the social and cultural life of the Vedic people and reflects their pastoral and agrarian lifestyle, social structures, and the importance of kinship and community. The

Rig Veda's hymns are foundational to Hindu theology, cosmology, and ritual practice. Its concepts and deities have evolved over millennia but continue to influence modern Hindu worship and philosophy.

Some hymns of the Rig Veda, such as the *Nasadiya Sukta* (the Hymn of Creation), delve into profound philosophical questions about the origin of the universe and the nature of existence. These early philosophical musings laid the groundwork for later Indian philosophical traditions. As the earliest known composition in an Indo-European language, the Rig Veda is a crucial text for linguists studying the development of these languages. Its poetic structure and sophisticated use of metaphor and symbolism also highlight the literary achievements of the Vedic age.

Here are some example texts and explanations from the Rig Veda*:

**Please note that in the phonetic representation, long vowels are indicated with a macron (ā), and retroflex sounds are represented with a dot below the letter (ṁ).*

Hymns to Deities:

The Rig Veda contains hymns dedicated to various deities such as Agni (fire), Indra (thunder), Varuna (the cosmic order), Surya (the sun), and Vishnu (preserver of the universe). These hymns praise the attributes, powers, and manifestations of the gods, seeking their blessings, protection, and favor.

Example Text (Rig Veda 1.1.1):

अग्‌नमिीळे पुरोहतिं यज्‌ञस्‌य देवं ऋत्‌वजिम्।

होतारं रत्नधातमम् ॥

Agnimīḷe purohitaṁ yajñasya devaṁ ṛtvijam ।

Hotāraṁ ratnadhatamam ॥

Translation:

"I Laud Agni, the chosen Priest, God, minister of sacrifice, The hotar, lavishest of wealth."

Explanation: This hymn praises Agni, the fire deity, as the chosen priest and mediator between humans and gods in sacrificial rituals. Agni is revered for bestowing wealth and prosperity upon the worshiper, symbolizing the transformative power of fire in rituals and spiritual practice.

Invocation and Prayer: Many hymns in the Rig Veda serve as invocations and prayers, invoking the gods for protection, prosperity, and guidance. These hymns express the devotion and reverence of the ancient seers towards the divine forces and seek their benevolence.

Example Text (Rig Veda 1.89.1):

इन्द्रं मित्रं वरुणमग्निमाहुरथो दिव्यः स सुपर्णो गरुत्मान ।

एकं सद विप्रा बहुधा वदन्त्यग्निं यमं मातरिश्वानमाहुः ॥

Indraṁ mitraṁ varuṇamagnimāhuratho divyaḥ sa suparṇo garutmān ।

Ekam sad viprā bahudhā vadantyagniṁ yamaṁ mātariśvānamāhuḥ ॥

Translation:

"They call him Indra, Mitra, Varuna, Agni, and he is heavenly nobly-winged Garutman.

To what is One, sages give many a title they call it Agni, Yama, Matarishvan."

Explanation: This verse highlights the multifaceted nature of the divine, represented by various names and forms such as Indra, Mitra, Varuna, and Agni. It underscores the unity underlying diversity and emphasizes the interconnectedness of all gods as manifestations of the supreme reality.

Cosmic Order and Natural Phenomena: The Rig Veda reflects a deep understanding and reverence for the natural world. It portrays the universe as governed by a divine order (rita) and recognizes the interconnectedness of all living beings with the cosmic forces.

Example Text (Rig Veda 1.164.46):

इन्द्रो मत्रिो वरुणो अग्नमिाहुरथो दवि्यः स सुपर्णो गरुत्मान ।

एकं सद वपि्रा बहुधा वदन्त्यग्नर्ि यमं मातरशि्वानमाहुः ॥

Indro mitro varuṇo agnimāhuratho divyaḥ sa suparṇo garutmān ।

Ekam sad viprā bahudhā vadantyagniṁ yamaṁ mātariśvānamāhuḥ ॥

Translation:

"Indra, Mitra, Varuna, and Agni these bear the name of gods, and heavenly-winged Garutman.

To what is One, sages give many a title they call it Agni, Yama, Matarishvan."

Explanation: This verse reiterates the diverse names and forms attributed to the divine, emphasizing the unity of the gods despite their different appellations.

It reflects the ancient Vedic understanding of the divine as both immanent and transcendent, pervading all aspects of creation.

Creation Myth and Cosmology: The Rig Veda contains hymns that describe the origins of the universe and the cosmic order. These hymns often depict creation as emerging from the primordial void and the divine sacrifice of the gods.

Example Text (Rig Veda 10.129.1-7):

नासदासीन्नो सदासीत्तदानी नासीद्रजो नो व्योमा परो यत् ।

कमिावरीवः कुह कस्य शर्मन्नम्भः कमिासीद्गहनं गभीरम् ॥

Nāsadāsīnno sadāsīttadānīṁ nāsīdrajo no vyomā paro yat ।

Kimāvarīvaḥ kuha kasya śarmanambhaḥ kimāsīdgahanam gabhīram ॥

Translation:

"Then even nothingness was not, nor existence,

There was no air then, nor the heavens beyond it.

What covered it? Where was it? In whose keeping?

Was there then cosmic water, in depths unfathomed?"

Explanation: This hymn explores the mystery of creation, contemplating the origin of the universe and the fundamental elements of existence. It reflects on the primordial void before the emergence of being and the profound questions surrounding the nature of reality and existence.

Moral and Ethical Teachings: Alongside the hymns of praise and invocations, the Rig Veda contains verses that impart moral and ethical guidance, emphasizing virtues such as truthfulness, righteousness, and compassion.

Example Text (Rig Veda 10.117.1):

अन्धं तमः प्रवशिन्ति ये ऽवदि्यामुपासते ।

ततो भूय इव ते तमो य उ वदि्यायां रताः ॥

Andhaṁ tamaḥ praviśanti ye 'vidyāmupāsate ।

Tato bhūya iva te tamo ya u vidyāyāṁ ratāḥ ॥

Translation:

"Blind darkness they enter who worship ignorance, into still greater darkness who delight in knowledge."

Explanation: This verse contrasts the paths of ignorance and knowledge, highlighting the consequences of each. It underscores the importance of seeking wisdom and enlightenment over succumbing to ignorance, emphasizing the value of spiritual insight in navigating life's journey.

Desire for Knowledge and Enlightenment: Many hymns express the desire for wisdom, enlightenment, and liberation from ignorance. They reflect the quest for deeper understanding and realization of ultimate truth.

Example Text (Rig Veda 1.164.6):

द्यौर्मीवाध्या तसि्रः प्रथमा नाभिः पृथवि्याः प्रथमो वरेण्यः ।

अद्य दवि्या द्यामर्णवाः सन्त एता वः शविा सन्तु नः ॥

Dyaurmīvādhyā tisraḥ prathamā nābhiḥ pṛthivyāḥ prathamo vareṇyaḥ ।

Adya divyā dyāmarṇavāḥ santa etā vaḥ śivā santu naḥ ॥

Translation:

> *"Two Birds with fair wings, knit with bonds of friendship, in the same sheltering tree have found a refuge.*
>
> *One of the twain eats the sweet Fig-tree's fruitage; the other eating not regardeth only."*
>
> Explanation: This metaphorical hymn depicts the human condition as two birds perched on the same tree, representing the individual soul (jivātman) and the supreme soul (paramātman). It illustrates the interplay between worldly desires and spiritual detachment, urging the seeker to transcend mundane distractions and realize their true nature.

These examples illustrate the diverse themes and rich philosophical insights present in the Rig Veda, making it a profound and multifaceted scripture of ancient Indian wisdom and spirituality.

2. Sama Veda

The Sama Veda is one of the four Vedas, intimately connected with the Rig Veda. It primarily comprises melodies (saman) and chants derived from the hymns of the Rig Veda, reorganized specifically for ritualistic chanting during Vedic sacrifices and ceremonies. The Sama Veda holds a unique place in Vedic literature and religious practice, emphasizing the musical and liturgical aspects of Vedic worship.

The text consists of about 1,875 verses, most of which are taken from the Rig Veda, particularly from its eighth and ninth mandalas. These verses are set to melodies and arranged in a way that facilitates their use in specific rituals and sacrificial ceremonies. The Sama Veda is divided into two main parts: the *Archika* (which contains the hymns) and the *Gana* (which includes the melodies). The Gana further divides into the *Gramageya* (sung in villages) and the *Aranyageya* (sung in the forests).

It is primarily a liturgical text used by the Udgatr priests during the Soma sacrifice and other Vedic rituals. The melodious chanting of the Sama Veda is believed to invoke divine forces, creating a sacred atmosphere conducive to the success of the rituals. By this token, the Sama Veda is the foundation of Indian classical music. The specific melodies and rhythmic patterns laid out in it have influenced the development of both Hindustani and Carnatic music traditions. The practice of singing these verses underscores the integral role of music in Vedic culture and the very act of chanting the Sama Veda is considered an act of devotion and meditation. The vibrations created by the precise intonation of the chants are believed to have a profound spiritual impact, both on the chanter and the listeners, promoting mental clarity and spiritual upliftment.

The Sama Veda also showcases the artistic and poetic reworking of the Rig Veda hymns into musical compositions. This transformation highlights the creative and adaptive nature of Vedic literature, showing how sacred texts were preserved and utilized in different forms. While the Rig Veda serves as the source of the hymns, the Sama Veda focuses on their musical rendition. This illustrates the interconnectedness of the Vedic texts and their comprehensive approach to spiritual and ritual practices. The Sama Veda's reliance on the Rig Veda's hymns demonstrates the collaborative and cumulative nature of the Vedic canon.

Melodies and Chants: The Sama Veda is characterized by its musical aspect, with verses from the Rig Veda rearranged into various musical patterns known as samans. These chants are sung by priests during Vedic rituals, accompanied by specific melodies and rhythms.

Example Text (Sama Veda 1.1.1):

अग्ने यं यज्ञमध्वरं विश्वतः परिभूरसि।

स इद्देवेषु गच्छति॥

Agne yaṁ yajñamadhvaraṁ viśvataḥ paribhūrasi ।

Sa iddeveṣu gacchati ॥

Translation:

"O Agni, the well-sung yajna (sacrifice) which you lead from all sides, goes to the gods."

Explanation: This verse from the Sama Veda praises Agni, the fire deity, and acknowledges the sacrificial rituals conducted with melodious chants led by Agni. It highlights the importance of Agni as the mediator between humans and gods in Vedic rituals, ensuring the offerings reach the divine realm.

Invocation and Devotion: Similar to the Rig Veda, the Sama Veda contains hymns and invocations to various deities, seeking their blessings and favor during rituals. These hymns express devotion and reverence towards the divine forces.

Example Text (Sama Veda 337):

या ते द॒क्षा शर्॒म या ते व॒श्याः।

या ते प्र॒जा अग्॒ने र॒यिं वरा॒ह्याः॥

Yā te dakṣā śarma yā te vaśyāḥ ।

Yā te prajā agne rayiṁ varāhyāḥ ॥

Translation:

"O Agni, to you belong skill and protection, offspring and prosperity."

Explanation: This hymn extols Agni's virtues and attributes, recognizing him as the bestower of skills, protector of devotees, and provider of prosperity and progeny. It reflects the devotion and reverence of the worshippers towards Agni, seeking his blessings for various aspects of life.

Integration with Rituals: Unlike the Rig Veda, which focuses more on philosophical and cosmological themes, the Sama Veda is closely associated with rituals and sacrifices. Its chants and melodies are specifically designed to accompany various stages of Vedic rituals, enhancing their efficacy and significance.

Example Text (Sama Veda 594):

अग्॒ने यमाय सोमपीतये।

स्॒वाहा॥

Agnē yamāya somapītaye ।

Swāhā ॥

Translation:

"O Agni, for the sake of Yama and Soma, hail!"

Explanation: This chant invokes Agni, addressing him as the intermediary between Yama (the god of death) and Soma (the sacred drink), essential components of Vedic rituals. It illustrates the integration of Agni's presence in facilitating communication between humans and the divine realm during rituals.

Aesthetic and Musical Emphasis: The Sama Veda places a strong emphasis on aesthetics and musicality. Its verses are arranged in a manner suitable for chanting, with attention to meter, rhythm, and melody, creating a sublime and harmonious atmosphere during rituals.

Example Text (Sama Veda 936):

तत्त्वं सदस्योपास्ते ।

स्वाहा ॥

Tattvaṁ sadasyopāste ।

Swāhā ॥

Translation:

"That truth he invokes who sacrifices."

Explanation: This verse underscores the significance of truth (satya) in Vedic rituals, emphasizing that those who perform sacrifices invoke the ultimate truth. It reflects the belief that rituals conducted with sincerity and adherence to truth lead to spiritual merit and divine favor.

Praise of Divine Attributes: Like other Vedas, the Sama Veda contains hymns praising the virtues and attributes of various deities, highlighting their greatness and significance in the cosmic order.

Example Text (Sama Veda 164):

अग्निमीळे पुरोहितिं यज्ञस्य देवं ऋत्विजिम् ।

होतारं रत्नधातमम् ॥

Agnimīḷe purohitaṁ yajñasya devaṁ ṛtvijam ।

Hotāraṁ ratnadhatamam ॥

Translation:

"I Laud Agni, the chosen Priest, God, minister of sacrifice, The hotar, lavishest of wealth."

Explanation: This hymn praises Agni for his pivotal role as the chief priest in Vedic rituals, facilitating communication between humans and gods. It acknowledges Agni's generosity in bestowing wealth and prosperity upon the worshippers, signifying the importance of his blessings in rituals.

Invocation for Blessings and Prosperity: Many hymns in the Sama Veda are invocations seeking divine blessings for prosperity, well-being, and fulfillment of desires.

Example Text (Sama Veda 167):

आ देवो देवमनि्द्रं हवामहे पुरोहितिं सोमं तस्थुः ।

सविता देवं सवितुर्धियिं यामः स्वस्तिनो गृह्यताम् ॥

Ā devo devamindraṁ havāmahe purohitaṁ somaṁ tasthuḥ ।

Savitā devaṁ savitur dhiyaṁ yāmaḥ svasti no gṛhyatām ॥

Translation:

"Come, O Gods, to the God Indra we call, to the Priest, Soma, stand thou by; O Savitar, to Savitar's holy power, to prosperity may we come."

Explanation: This invocation calls upon various gods, including Indra, Soma, and Savitar, seeking their presence and blessings during the ritual. It expresses the devotee's aspiration for divine assistance and prosperity, highlighting the importance of invoking multiple deities for different aspects of life.

These examples illustrate the diverse themes and spiritual significance found within the Sama Veda, enriching the understanding of its nature and contents.

3. Yajur Veda

The Yajur Veda is one of the four Vedas, primarily consisting of prose mantras and verses used by priests during Vedic rituals, particularly sacrificial ceremonies. It is a practical guide for conducting rituals, providing detailed instructions and formulae for the performance of yajnas (sacrifices) and other sacred rites. The Yajur Veda is divided into two major parts: the *Shukla* (White) Yajur Veda and the *Krishna* (Black) Yajur Veda, each with its own samhitas (collections).

Unlike the Rig Veda, which is composed mainly of hymns, the Yajur Veda includes a combination of prose mantras and verses. These are designed to be recited by the Adhvaryu priests, who play a crucial role in performing the rituals. The Yajur Veda is divided into the Shukla Yajur Veda and Krishna Yajur Veda. The Shukla Yajur Veda is more systematic and organized, while the Krishna Yajur Veda contains a mix of prose and commentary, with explanations interwoven with the mantras.

The Yajur Veda is essential for the correct performance of Vedic rituals. It provides the specific formulae and procedures required for various sacrifices, including the *Agnihotra* (fire sacrifice), *Ashvamedha* (horse sacrifice), and *Rajasuya* (royal consecration). The precise recitation of these mantras is believed to ensure the efficacy and success of the rituals. The texts serve as a practical manual for priests, detailing every aspect of ritual performance, from the preparation of the sacrificial ground to the offerings made to the deities. They ensure that the rituals are conducted in accordance with ancient traditions and prescribed norms.

The mantras and verses of the Yajur Veda are considered powerful tools for invoking divine forces and seeking blessings. The rituals guided by these texts are seen as means of maintaining cosmic order (rta) and ensuring harmony between humans and the divine. Its prose mantras and ritualistic verses contribute to the richness of Vedic literature. They offer insights into the evolution of Vedic language and the use of ritualistic expressions in ancient India. Equally, the Yajur Veda complements the Rig Veda and Sama Veda by focusing on the procedural aspects of rituals. While the Rig Veda provides the hymns and the Sama Veda the melodies, the Yajur Veda details the actions and recitations required to perform the rituals effectively.

Performance of Sacrificial Rituals: The Yajur Veda contains detailed instructions and mantras used by priests during sacrificial ceremonies, guiding every aspect of the ritual, from the preparation of the altar to the invocation of various deities.

Example Text (Yajur Veda Taittiriya Samhita 1.1.1):

अग्निमीळे पुरोहितं यज्ञस्य देवं ऋत्विजम्।

होतारं रत्नधातमम्॥

Agnimīḷe purohitaṁ yajñasya devaṁ ṛtvijam।

Hotāraṁ ratnadhatamam॥

Translation:

"I Laud Agni, the chosen Priest, God, minister of sacrifice, The hotar, lavishest of wealth."

Explanation: This verse praises Agni, the fire deity, as the central figure in sacrificial rituals. Agni is acknowledged as the chosen priest who facilitates communication between the human and divine realms, ensuring the success and abundance of the sacrifice.

Invocation for Divine Blessings: Like other Vedas, the Yajur Veda contains hymns and invocations calling upon various deities for blessings, protection, and guidance during rituals.

Example Text (Yajur Veda Taittiriya Samhita 1.5.1):

तत्सवितुर्वरेण्यं भर्गो देवस्य धीमहि।

धियो यो नः प्रचोदयात्॥

Tatsaviturvareṇyaṁ bhargo devasya dhīmahi।

Dhiyo yo naḥ pracodayāt॥

Translation:

"May we attain that excellent glory of Savitar the God: So may he stimulate our prayers."

Explanation: This hymn invokes Savitar, the solar deity, seeking his divine grace and inspiration to elevate the prayers of the worshippers. It expresses the aspirational desire to attain the divine radiance and blessings bestowed by Savitar.

Preservation of Cosmic Order: The Yajur Veda emphasizes the importance of maintaining the cosmic order (rita) through righteous conduct and adherence to ethical principles.

Example Text (Yajur Veda Taittiriya Samhita 1.8.6):

तस्मादयत्रेमा आत्मेदा आत्मानं परेद्दुः।

न वशि्वेषां प्रयिं भवतिा॥

Tasmādyatrema ātmedā ātmānaṁ pared duḥ ।

Na viśveṣāṁ priyaṁ bhavati ॥

Translation:

"Therefore, where these beings, having regarded Self as the Self of the other, do not strike or kill one another, there they are dear to each other."

Explanation: This verse emphasizes the importance of non-violence and mutual respect in preserving the cosmic order (rita) and fostering harmonious relationships within society. It underscores the ethical principle of treating others with compassion and recognizing the interconnectedness of all beings.

Chants for Purification: The Yajur Veda includes mantras for purifying the mind, body, and environment, ensuring the sanctity and efficacy of the rituals being performed.

Example Text (Yajur Veda Taittiriya Samhita 1.1.1):

अपो नाराः सुसंवदति।

Apo nārāḥ susaṁvadati ।

Translation:

"The waters purify."

Explanation: This mantra highlights the purifying power of water, a vital element used in Vedic rituals for cleansing the mind, body, and environment. It signifies the importance of purity and sanctity in the performance of sacrificial ceremonies.

Invoking Deities for Prosperity: Many hymns in the Yajur Veda invoke deities for prosperity, wealth, and abundance, seeking their benevolence for the welfare of the sacrificer and the community.

Example Text (Yajur Veda Taittiriya Samhita 1.5.1):

इदं देवानाममृतं पपतिानं सोमस्याभषिग्।

Idaṁ devānāmamṛtaṁ papitānaṁ somasyābhiṣag।

Translation:

"This is the invigorating drink of the gods, the immortal, well-drunk Soma."

Explanation: This hymn extols Soma, the sacred drink of the gods, as a source of vitality, immortality, and divine favor. It invokes blessings for prosperity and abundance, symbolizing the nourishing and rejuvenating essence of Soma.

Guidance for Righteous Living: Alongside rituals, the Yajur Veda imparts teachings on righteous conduct, ethical behavior, and the importance of fulfilling one's duties in society.

Example Text (Yajur Veda Taittiriya Samhita 1.8.6):

अनेन प्रयत्नेन पुनरागच्छती।

Anena prayatnena punarāgacchati।

Translation:

"By this exertion one goes again."

Explanation: This verse underscores the significance of righteous effort and ethical conduct in shaping one's destiny. It conveys the idea that through diligent effort and virtuous actions, individuals can attain spiritual evolution and progress towards higher states of existence.

The Yajur Veda's focus on prose mantras and ritual instructions makes it indispensable for the performance of Vedic ceremonies. Its detailed guidance ensures the proper conduct of rituals, reflecting the Vedic emphasis on precision and reverence in worship. The Yajur Veda's contributions to Vedic literature and its practical application in rituals underscore its enduring significance in the religious and cultural life of ancient and modern Hinduism.

4. Atharva Veda

The Atharva Veda is the fourth Veda and is distinct from the other three in its focus on rituals, incantations, and magic spells, along with hymns and prayers. It encompasses a wide range of topics, including healing, protection, and everyday life, reflecting the practical concerns and spiritual needs of ancient Indian society. The Atharva Veda provides a unique glimpse into the mystical and ritualistic traditions of the Vedic period.

It contains numerous spells and incantations aimed at curing diseases, warding off evil spirits, and ensuring success in various endeavors. These rituals address both mundane and spiritual aspects of life. Like the other Vedas, the Atharva Veda includes hymns and prayers dedicated to various deities. However, these often have a more practical and immediate focus, relating to health, prosperity, and protection. Divided into 20 books (kandas), the Atharva Veda contains about 730 hymns with over 6,000 mantras. These are further categorized into charms and spells (Brahma Veda) and hymns of a more spiritual nature (Angiras Veda).

One of the notable features of the Atharva Veda is its focus on healing and medicine. It includes numerous herbal remedies, prayers for health, and incantations to cure ailments, reflecting early medical knowledge and practices. It also contains spells and charms designed to protect individuals from harm, ensure success in various endeavors, and bring prosperity. These rituals often involve invoking divine forces for support and guidance, touching on mystical and spiritual themes, with prayers and hymns that seek to understand and connect with the divine forces governing the universe.

Incantations for Healing and Protection: The Atharva Veda contains numerous incantations for healing ailments, warding off evil spirits, and ensuring protection from malevolent forces.

Example Text (Atharva Veda 1.6.1):

अपं नपातु सूर्यः।

Apaṁ napātu sūryaḥ।

Translation:

“May the sun god protect our waters.”

Explanation: This verse invokes the sun god to safeguard the purity and vitality of water, highlighting the importance of water as a life-sustaining element. It reflects the belief in the divine power to protect natural resources and ensure their abundance for sustenance and well-being.

Charms for Prosperity and Success: Like other Vedas, the Atharva Veda includes charms and rituals aimed at attracting prosperity, wealth, and success in various endeavors.

Example Text (Atharva Veda 2.1.1):

योऽस्मत्पुरोधां वेत्ति।

Yo'smatpurodhāṁ vetti ।

Translation:

"He who knows our abode."

Explanation: This charm invokes blessings for prosperity and success, acknowledging the importance of divine knowledge and protection in achieving one's goals and aspirations. It signifies the desire for divine guidance and support in navigating life's journey towards abundance and fulfillment.

Mantras for Warding off Diseases: Atharva Veda contains mantras and rituals for averting diseases and ensuring well-being, both physical and mental.

Example Text (Atharva Veda 4.15.1):

अश्वासो मा विन्दति।

Aśvāso mā vindati ।

Translation:

"May he not find breath."

Explanation: This mantra aims to ward off diseases and afflictions, invoking divine protection against threats to physical well-being. It expresses the fervent desire for health and vitality, seeking divine intervention to prevent illness and ensure the continued well-being of the individual.

Invocation for Strength and Courage: There are invocations in the Atharva Veda aimed at invoking strength, courage, and valor to overcome obstacles and challenges.

Example Text (Atharva Veda 7.2.1):

तत्ते मद्वर्धनं कुरु ।

Tatte madvardhanaṁ kuru ।

Translation:

"May you increase in might."

Explanation: This invocation seeks to imbue the individual with strength, courage, and valor to overcome challenges and obstacles. It reflects the aspiration for personal growth and empowerment, invoking divine blessings for inner resilience and fortitude in the face of adversity.

Rituals for Household Harmony: The Atharva Veda provides rituals and prayers for maintaining harmony and well-being within the household and family.

Example Text (Atharva Veda 14.2.1):

तान्नो मुञ्चामि।

Tānno muñcami ।

Translation:

"I release them."

Explanation: This prayer aims to resolve conflicts and restore harmony within the household or family. By releasing tensions and grievances, it seeks to promote mutual understanding and reconciliation, fostering a sense of unity and well-being among family members.

Invocation of Deities for Blessings: Similar to other Vedas, the Atharva Veda invokes various deities for blessings, prosperity, and divine assistance.

Example Text (Atharva Veda 11.5.1):

स वै न आवविश देवतायाः ।

Sa vai na āviveśa devatāyāḥ ।

Translation:

"He has entered the deity."

Explanation: This verse acknowledges the presence and influence of the deity invoked in the ritual, affirming the divine connection established through worship. It signifies the belief in the efficacy of rituals in attracting divine blessings and assistance, ensuring prosperity, and spiritual fulfillment.

These examples highlight the diverse content of the Atharva Veda, which encompasses a wide range of rituals, spells, prayers, and incantations aimed at addressing various aspects of life and spirituality. Its content reflects a comprehensive approach to addressing both the spiritual and material needs of the Vedic people. The Atharva Veda's contributions to Vedic literature, its focus on healing and protection, and its insights into the daily life and beliefs of ancient India underscore its unique and enduring significance.

This was a quick and perhaps a deep enough immersion in the Vedas for most people. Now, let us move on to find ourselves amongst the contents of the Upanishads!

Do it yourself

Duli, a scientist, was speaking to Katja, a doctor, or more accurately venting his frustrations to her. He had just restarted his study of the Vedas after a gap of a decade, which he had spent studying sciences. His intent now was to weave more Vedic content in his own spiritual workshops, which he actively engaged in as a volunteer activity. He was complaining about how badly some of the Vedas were translated into English, and how the commentaries of the Upanishads were 'literal' translations that lacked the context of why they said what they said, and were present in the Vedas they were part of.

Katja patiently heard him out and then asked him if he believed that teaching the correct interpretation of Vedas and Upanishads was an act of Dharma. *Duli responded, "It's probably the greatest Dharma for a teacher.*" Katja followed up, asking him how he knew so much about the gaps in translations, even though he had not studied them as deeply as some of the Indology scholars. Duli responded by saying he was guided by an inner knowingness, *an invisible hand.*

Katja then asked him why he didn't become a full-time teacher himself, given that he was guided by Dharma and believed that teaching the Vedas and Upanishads was a sacred duty, especially since he was deeply concerned about the lack of good teachers and quality content.

Duli went quiet for a few moments, before replying, "*I don't know. Maybe I have a fear I have not transcended.*"

Kajta asked him if that was just Maya showing up as his fear.

Today, Duli is a well-known spiritual teacher and he and Katja still collaborate. Duli will soon be completing what he calls his "most inspired work" – a 21st century revisit of the Vedanta!

"The Upanishads are the only books in the world which show the essential nature of human beings and the human mind."

– Carl Jung

CHAPTER 3

A JOURNEY THROUGH THE 10 MAIN UPANISHADS

The Upanishads, considered to be the culmination of Vedic wisdom, are a collection of more than 200 philosophical Sanskrit texts within Hinduism that explore profound metaphysical and spiritual concepts. They are traditionally considered to be Shruti scriptures: divinely revealed truths heard by ancient sages during deep meditation and contemplation. Characterized by their poetic and philosophical nature, they delve into the deepest mysteries of existence, offering insights into the nature of reality, the self, and the ultimate truth (Brahman). You will find the list of these Upanishads in the appendix.

The Sanskrit word "Upanishad" consists of two parts: "upa" and "nishad." "Upa" means "near" or "close to," and "nishad" means "to sit down." Therefore, the literal meaning of "Upanishad" is "sitting down near" or "sitting down closely."

In a spiritual context, the term "Upanishad" refers to the ancient Indian texts made up of philosophical and spiritual teachings imparted by a guru (teacher) to their students in an intimate, close setting. These teachings often involve profound discussions on topics such as the nature of reality, the self (atman), Brahman (the ultimate reality), karma (action), dharma (duty/righteousness), and moksha (liberation).

The Upanishads, considered to be the apex of Vedic thought, are also known as Vedanta, signifying "the end of the Vedas" or "the essence of the Vedas." They hold a revered status in Hindu philosophy, serving as foundational texts for numerous schools of Vedanta and exerting significant influence on other Indian philosophical traditions.

Positioned as the final section or "anta" of the Vedas, succeeding the Samhitas, Brahmanas, and Aranyakas, the Upanishads mark a transition from the external rituals and sacrificial practices of earlier Vedic literature to a deeper exploration of metaphysical realities.

Utilizing dialogues, allegories, and parables, they impart profound philosophical insights and spiritual wisdom. Their teachings not only shape Hindu thought but also contribute to the evolution of diverse philosophical and spiritual traditions across India.

The terms Vedanta and Upanishads are often used interchangeably, even among scholars, highlighting the close connection between the two. Vedanta expounds upon the metaphysical and spiritual principles elucidated in the Upanishads, providing interpretations, commentaries, and philosophical systems based on these ancient texts.

The teachings of Vedanta encompass various schools of thought, including Advaita (non-dualism), Dvaita (dualism), and Vishishtadvaita (qualified non-dualism), among others. These schools offer different perspectives on the nature of reality, the self, and the ultimate goal of human existence, drawing inspiration and guidance from the profound wisdom found in the Upanishads.

Thus, Vedanta serves as a philosophical framework that elucidates, expands upon, and applies the teachings of the Upanishads, offering profound insights into the nature of existence and the path to spiritual realization.

Some of the Upanishads are considered more important and widely studied than others. Traditionally, there are ten principal Upanishads that are often referred to as the "major" or "principal" Upanishads. I have listed them below, along with the names of Vedas they are found in, and a brief description of their content follows.

1. Īśā - Yajur Veda
2. Kena - Sama Veda
3. Kaṭha - Yajur Veda
4. Praśna - Atharva Veda
5. Muṇḍaka - Atharva Veda
6. Māṇḍūkya - Atharva Veda
7. Taittirīya - Yajur Veda
8. Aitareya - Rig Veda
9. Chāndogya – Sama Veda
10. Bṛhadāraṇyaka - Yajur Veda

Let's now immerse ourselves in some of their contents and themes!

1. Īśā Upanishad - *Yajur Veda*

The Isa Upanishad, also known as the Isavasya Upanishad, is one of the shortest and most profound Upanishads, belonging to the Shukla Yajur Veda. Despite its brevity, it encapsulates profound philosophical and spiritual insights into the nature of existence, the relationship between the individual soul and the universal consciousness, and the path to spiritual liberation.

Here are some example texts from the Īśā Upanishad, along with explanations:

Verse 1 (Mantra 1):

ईशावास्‌यमदिं सर्‌वं यत्‌कञ्‌चि जगत्‌यां जगत्‌।

तेन त्‌यक्‌तेन भुञ्‌जीथा मा गृधः कस्‌यस्‌वद्‌िधनम्‌॥

Ishāvāsyamidaṁ sarvaṁ yatkiñca jagatyāṁ jagat ।

Tena tyaktena bhuñjīthā mā gṛdhaḥ kasyasviddhanam ॥

Translation:

"The entire universe, whatever there is that is animate or inanimate, is pervaded by the Lord (Īśa). Enjoy (the fruits of) renunciation. Do not covet what belongs to others."

Explanation:

This verse highlights the omnipresence of the divine (Īśa) in the universe. It teaches detachment (tyāga) from material possessions and urges individuals to enjoy the world without greed or attachment.

Verse 15 (Mantra 15):

हरिण्‌मयेन पात्‌रेण सत्‌यस्‌यापहिितं मुखम्‌।

तत्त्वं पूषन्नपावृणु सत्यधर्माय दृष्टये ॥

Hiraṇmayena pātreṇa satyasyāpihitaṁ mukham ।

Tattvaṁ pūṣannapāvṛṇu satyadharmāya dṛṣṭaye ॥

Translation:

"Through a golden vessel, the face of truth is hidden. O Pushan, remove that, for the sake of the Truth, which lies under the concealment of gold, so that I, who am devoted to truth, may behold it."

Explanation:

This verse uses symbolic language to convey the idea that the ultimate truth (satya) is often concealed by material possessions or distractions. The seeker prays to the divine deity Pushan to remove the veil of ignorance and reveal the truth.

Verse 17 (Mantra 17):

अनेजदेकं मनसो जवीयो नैनधेवा आप्नुवन्पुर्वमर्षत् ।

तद्धावतोऽन्यानत्येत तष्ठित् तस्मिन्निपो मातरिश्वा दधाति॥

Anejadekaṁ manaso javīyo nainadhevā āpnuvanpurvamarṣat ।

Taddhāvato'nyānatyetatiṣṭhat tasminnapo mātarishvā dadhāti ॥

Translation:

"Unmoving, it is faster than the mind. The senses could not overtake It, since It had run ahead. Remaining stationary, It outruns all other runners. Being there, Matarishva supports all activities."

Explanation:

This verse elucidates the omnipresence and omnipotence of the divine reality (Brahman), which transcends the limitations of the mind and the senses. It illustrates the ineffable nature of the ultimate reality that sustains and supports all existence.

These examples provide a glimpse into the profound philosophical and spiritual teachings of the Īśā Upanishad, which continue to inspire seekers on the path of self-realization and spiritual enlightenment. Its concise and profound teachings provide a comprehensive guide to understanding the nature of existence, while emphasizing the unity of the self with the universal consciousness, ethical living, and the balance between action and renunciation.

2. Kena Upanishad - Sama Veda

The Kena Upanishad explores the nature of Brahman and the relationship between the individual self (Atman) and the ultimate reality. It teaches that Brahman cannot be known through the senses but can be realized through spiritual intuition and insight. Named after the first word of its opening verse, "Kena," meaning "by whom," the Kena Upanishad delves into the question of the ultimate cause or power behind all phenomena, guiding the seeker on a journey of self-inquiry and realization.

This Upanishad consists of four sections, containing a mix of prose and verse, and presents a dialogue between a student and a teacher touching on metaphysical questions about the nature of reality and consciousness. The Kena Upanishad focuses on the ineffable nature of Brahman, emphasizing that it is beyond sensory perception and intellectual understanding. It also delves into the relationship between the individual self (Atman) and Brahman, teaching that the true self is not separate from the ultimate reality but is a manifestation of it.

The text also provides profound mystical insights into the nature of consciousness and existence. It uses paradoxes and metaphors to illustrate that Brahman is the essence of all things yet beyond all description. Philosophical inquiry and introspection are encouraged, with seekers prompted to question the nature of their own consciousness and the ultimate reality, fostering a deep sense of wonder and contemplation.

Here are some example texts from the Kena Upanishad, along with explanations:

Kena Upanishad 1.1:

नम ऋषभि्यो मन्त्रभ्यः ।

नम आप्याय चक्षुषे ।

नम ध्येयाय च पश्यते ।

नमो दवि्याय तेजसे ।

Nama ṛṣibhyo mantrabhyaḥ ।

Nama āpyāya cakṣuṣe ।

Nama dhyeyāya ca paśyate ।

Namo divyāya teyase ।

Translation:

"Salutations to the Rishis, salutations to the mantras, salutations to the Devatas that bestow

eyesight, salutations to the one who is meditated upon, salutations to the divine effulgence."

Explanation:

This opening verse of the Kena Upanishad pays homage to the Rishis (sages), the mantras (sacred chants), and the divine forces that grant perception and insight. It sets the tone for the Upanishads' exploration of the spiritual path and the pursuit of transcendental knowledge.

The Kena Upanishad further delves into the nature of Brahman through a dialogue between a teacher and a disciple, exploring profound metaphysical concepts such as the nature of perception, consciousness, and the ultimate reality beyond the senses.

Kena Upanishad 1.4-5:

केनेषतिं पततप्रेषतिं मनः ।

केन प्राणः प्रथमः प्रैतयुक्तः ।

केनेषतिां वाचममिां वदन्ती

चक्षुः श्रोत्रं क उ देवो युनक्ती॥

Keneṣitaṁ patati preṣitaṁ manaḥ ।

Kena prāṇaḥ prathamaḥ praiti yuktaḥ ।

Keneṣitāṁ vācamimāṁ vadanti ।

Cakṣuḥ śrotraṁ ka u devo yunakti ॥

Translation:

"By whom directed does the mind proceed to its object?

By whom impelled does the first vital air move forward?

By whom directed do people speak this speech?

What god joins the eyes and the ears?"

Explanation:

These verses from the Kena Upanishad inquire into the ultimate cause behind various faculties such as the mind, breath, speech, sight, and hearing. They prompt introspection into the source of all action and perception, guiding the seeker to contemplate the underlying unity and consciousness that animate the phenomenal world.

Kena Upanishad 1.6:

यदि मन्यसे सुवेदेति दभ्रमेवापि नूनं वेदानुवच्छिेत्स्व तं केन वजिानीयात् ॥

Yadi manyase suvedeti dabhramevāpi nūnaṁ vedānuvic-chetsva taṁ kena vijānīyāt ॥

Translation:

"If you think, 'I know Brahman well,' then surely you know but a part of Brahman. Reflect on this. How can you claim to know the unknown?"

Explanation:

This verse challenges the notion of complete understanding or mastery of Brahman, the ultimate reality. It encourages seekers to contemplate the limitations of their knowledge and to inquire further into the nature of the unknown. It underscores the importance of humility and ongoing inquiry in the pursuit of spiritual knowledge.

Kena Upanishad 2.1:

यददिं कंचि जगत्सर्वं प्राणा एजत निहितिं वा एतद्यददिं कंचि जगत्सर्वं प्राणा एजत निहितिं वा एतद्यददिं कंचि जगत्सर्वं प्राणा एजत निहितिं वा एतद्यददिं कंचि जगत्सर्वं प्राणा एजत निहितिं वा एतद्यददिं कंचि जगत्सर्वं प्राणा एजत निहितिं वा एतद्यददिं कंचि जगत्सर्वं प्राणा एजत निहितिं वा एतद्यददिं कंचि जगत्सर्वं प्राणा एजत निहितिं वा एतद् ॥

Yadidaṁ kiṁ ca jagatsarvaṁ prāṇā ejati nihitaṁ vā etadyadidaṁ kiṁ ca jagatsarvaṁ prāṇā ejati nihitaṁ vā etadyadidaṁ kiṁ ca jagatsarvaṁ prāṇā ejati nihitaṁ vā etadyadidaṁ kiṁ ca jagatsarvaṁ prāṇā ejati nihitaṁ vā etad ॥

Translation:

"That which moves not with the mind, but by which the mind is moved, know that alone to be Brahman, the immortal. In it is established the activity of all."

Explanation:

This verse elucidates the concept of Brahman as the ultimate source of all activity and movement in the universe. It emphasizes that Brahman is not an object of the mind but the transcendent reality that animates all beings and phenomena. It prompts seekers to recognize the presence of Brahman as the underlying substratum of existence.

Kena Upanishad 3.1:

यद्वाचा नानुवदि्यते तदेव ब्रह्म त्वं वदि्धनिेदं यददिं उपासते ॥

Yadvācā nānuvidyate tadeva brahma tvaṁ viddhi nedaṃ yadidaṁ upāsate ॥

Translation:

"That which cannot be apprehended by speech but by which speech is apprehended, know that alone to be Brahman, not this that people worship."

Explanation:

This verse highlights the ineffable nature of Brahman, which transcends the limitations of speech and conceptualization. It suggests that Brahman can only be realized through direct experience or intuition, beyond the grasp of mere words or rituals. It encourages seekers to seek the true essence of Brahman rather than being confined to superficial forms of worship.

Overall, the Kena Upanishad elucidates profound spiritual truths and philosophical insights, guiding aspirants on the path to self-realization and union with Brahman, the ultimate reality. Its teachings highlight the limitations of sensory and intellectual knowledge and emphasize the need for direct spiritual insight. Through its poetic verses and philosophical dialogue, the Kena Upanishad guides seekers on a journey of self-inquiry, urging them to transcend ordinary perceptions and realize the true essence of existence.

3. Kaṭha Upanishad – Yajur Veda

The Katha Upanishad is a philosophical text found in the Yajur Veda, specifically within the Katha branch of the Yajur Veda. It is a profound dialogue between the sage Yama and the young seeker Naciketas, exploring themes of life, death, and the nature of the self. The text is divided into two chapters (adhyayas), each containing three sections (vallis). It comprises a total of 120 mantras, systematically discussing the path to self-realization.

The dialogue begins with Naciketas questioning Yama about the nature of death and what lies beyond. Yama's teachings provide deep insights into the immortality of the soul (Atman) and the nature of existence. A central theme of the Kaṭha Upanishad is the exploration of the true self (Atman), in which ilt emphasizes that the Atman is eternal, indestructible, and beyond physical death, urging seekers to look inward for true knowledge.

Yama also instructs Naciketas on the path to moksha (liberation), which involves self-discipline, meditation, and knowledge (jnana). As well as highlighting the importance of realizing one's true nature and transcending the material world, it also imparts ethical and moral lessons, encouraging individuals to pursue righteous living and spiritual wisdom over materialistic desires.

Here are some example texts from the Katha Upanishad, along with explanations:

Kaṭha Upanishad 1.2.23:

नैषा तर्केण मतरिापनेया

प्रोक्तान्येनैव सुज्ञानाय प्रेष्ठ ।

यां त्वमापः सत्यधृतरि्बतासि

त्वात्तो धर्मं चरतिुं नगिृह्णीषे ॥

Naiṣā tarkena matirāpaneyā

Proktānyenaiva sujñānāya preṣṭha ।

Yāṁ tvamāpaḥ satyadhṛtirbatāsi

Tvātto dharmaṃ carituṃ nigṛhṇīṣe ॥

Translation:

"This knowledge cannot be attained by reasoning. It comes only when taught by another who has realized it. O Nachiketas, you have been in the proximity of Death, yet you are a true aspirant. Therefore, I shall tell you the eternal truth. Choose the boon you desire."

Explanation:

This verse highlights the ineffability of spiritual knowledge, suggesting that it cannot be comprehended through mere intellectual reasoning but must be transmitted by a realized teacher. It acknowledges Naciketas' sincerity and readiness for spiritual wisdom, prompting Yama to impart the profound truths of existence.

Kaṭha Upanishad 2.1.1:

उत्तिष्ठत जाग्रत प्राप्य वरान्निबोधत

क्षुरस्य धारा निशिता दुरत्यया

दुर्गं पथस्तत्कवयो वदन्ती।

Uttiṣṭhata jāgrata prāpya varānnibodhata

Kṣurasya dhārā niśitā duratyayā

Durgaṁ pathastat kavayo vadanti ।

Translation:

"Arise, awake, and learn from the wise ones who are adept in the scriptures, for the path is as sharp as a razor's edge, difficult to cross and hard to tread upon."

Explanation:

This verse serves as a wake-up call, urging the seeker to rouse from ignorance and heed the guidance of enlightened teachers. It metaphorically compares the spiritual path to a sharp razor's edge, signifying its challenging nature and the need for careful navigation and discernment.

Kaṭha Upanishad 2.1.:

अन्यच्छ्रेयोऽन्यदुतैव प्रेयस्ते

उभेनानार्त्थेनार्त्थं तरेऽवृणीथा ।

भजत तद्धा योगत इत्युक्तो

ज्ञानेन तु तदज्ञानं येषां नाशतिं आत्मनः ।

Anyacchreyo'nyadutaiva preyaste

Ubhena'nārththenārthaṁ tare'vṛṇīthā ।

Bhajata taddhā yogata ityukto

Jñānena tu tadajñānaṁ yeṣāṁ nāśitaṁ ātmanaḥ ।

Translation:

"Separate are the desirable and the beneficial. An uncertain path lies between them. The wise man distinguishes them through knowledge and chooses the beneficial over the merely pleasant. The foolish one, on the other hand, chooses the merely pleasant due to his lack of understanding."

Explanation:

This verse contrasts between the transient pleasures (the pleasant) and the eternal good (the beneficial), emphasizing the importance of discernment in making choices. It underscores the role of wisdom in distinguishing between short-term gratification and long-term spiritual growth, guiding the seeker toward the path of true fulfillment.

Using metaphors and allegories to illustrate complex concepts, the Kaṭha Upanishads' teachings are foundational to Vedantic philosophy, with its wealth of spiritual insights continuing to inspire and illuminate the quest for self-realization and ultimate truth.

4. Praśna Upanishad - Atharva Veda

The Prasna Upanishad is a significant philosophical text found within the Atharva Veda. It consists of six questions posed by six different seekers to the sage Pippalada. His answers address various aspects of spiritual knowledge, including the nature of the self, the universe, and the ultimate reality.

The text is divided into six chapters (prasnas), each corresponding to one of the questions. This systematic approach allows for an in-depth exploration of key philosophical concepts, such as, the nature of the self, cosmic understanding, the significance of prana, and the right spiritual path It also includes discussions on meditation, devotion, and the pursuit of wisdom, encouraging a holistic approach to inquiry. By combining intellectual understanding with spiritual practice, it guides seekers on how to integrate knowledge and experience in their spiritual journey.

Here are some example texts from the Prasna Upanishad, along with explanations:

Praśna Upanishad 1.2:

तमहं ब्रह्माणं बृहन्तमनुप्रविश्य

ब्रह्मविद्यां वदामि यो यो देवेभ्यो ब्राह्मणेभ्यश्च यः प्रतिष्ठितिः ।

तस्य ह स्वाहान्वतेऽन्यत्र धर्मदात् प्रतिष्ठा ।

तदा प्रतिष्ठां प्राप्य स धर्मं विचिचार स्वाहेत्येवं विद्वानमुपासीत ॥

Tamahaṁ brahmāṇaṁ bṛhantamanupraviśya

Brahmavidyāṁ vadāmi yo yo devebhyo brāhmaṇebhyaśca yaḥ pratiṣṭhitaḥ ।

Tasya ha svāhānvate'nyatra dharmadāt pratiṣṭhā ।

Tadā pratiṣṭhāṁ prāpya sa dharmaṁ vicacāra svāhettyevaṁ vidvānamupāsīta ॥

Translation:

"I will speak of the knowledge of Brahman, the great, having approached Him. To those devas and Brahmins who are dedicated to the devas and Brahmins, who are established in the deities, a seat is given elsewhere. On obtaining that seat, the wise man reflects on Dharma. Such a wise man should meditate on Him as 'Swaha'."

Explanation:

This verse emphasizes the importance of attaining knowledge of Brahman, the ultimate reality, through dedicated practice and inquiry. It suggests that those who are established in devotion and righteousness attain a special seat, allowing them to reflect upon Dharma (righteous duty). It encourages seekers to contemplate upon the divine essence denoted by the term "Swaha."

Praśna Upanishad 2.1:

अन्नं ब्रह्मेति व्यजानात् अन्नाद्भवन्ति भूतानि।

परोऽन्नं तद्वर्षति।

तस्मादन्नं तद्ब्रह्म।

Annam brahmeti vyajānāt annādbhavanti bhūtāni ।

Paro'nnam tadvarṣati ।

Tasmādannam tadbrahma ।

Translation:

"Food is verily Brahman. From food, all beings are born. Having been born, by food they live. And into food, they merge at the end. Therefore, food is indeed Brahman."

Explanation:

This verse elucidates the concept of Brahman as immanent in the material world, particularly in food. It highlights the essential role of food in sustaining life and underscores the interconnectedness between physical nourishment and the spiritual essence. It invites contemplation on the divine presence pervading all aspects of existence.

Praśna Upanishad 3.1:

क्षुत्पिपासायामदितिर्देवतेत्युपास्यो वाव तमेव देवतं यो दिवा नैमित्तिकं तमुपास्ते रात्रिं प्रातर्दिवं वा निष्ठितो यो रात्रिं नैमित्तिकं तमुपास्ते दिवं वा रात्रिं योऽर्वाग्यत्येते सर्वे प्रत्यगात्मानं क्षुत्पिपासायामुपास्ते।

Kṣutpipāsāyāmaditirdevatetyupāsyo vāva tameva devataṃ yo divā naimittikaṃ tamupāste rātriṃ prātardivaṃ vā niṣṭhito yo rātriṃ naimittikaṃ tamupāste divaṃ vā rātriṃ yo'rvāgyate sarve pratyagātmānaṃ kṣutpipāsāyāmupāste ।

Translation:

"The Sun, verily, is the deity presiding over hunger and thirst. It is He whom one should worship. The one who is attached to the daytime, considering it as the primary cause, worships Him during the night or in the early morning, while the one attached to the night worships Him during the day or at dusk. All these worship the individual Self, thinking it to be afflicted by hunger and thirst."

Explanation:

This verse elucidates the worship of the Sun as the deity presiding over hunger and thirst. It illustrates how individuals, driven by their attachment to either daytime or nighttime, worship the Sun during different periods. However, it ultimately reveals that all such worship is directed toward the individual Self, recognizing its essential nature and its perceived afflictions of hunger and thirst.

The Prasna Upanishad offers a rich and diverse exploration of spiritual knowledge through its question-and-answer format. Its teachings on the nature of the self, the cosmos, prana, and the path to liberation provide profound insights and guidance for seekers. The holistic approach it has to philosophical inquiry and spiritual practice make it an invaluable text in the Vedic tradition.

5. Muṇḍaka - Atharva Veda

The Mundaka Upanishad is a profound philosophical text found within the Atharva Veda. It consists of three chapters and discusses the pursuit of spiritual knowledge (vidya) and the path to self-realization. It distinguishes between lower knowledge (apara-vidya), which deals with worldly matters, and higher knowledge (para-vidya), which leads to liberation.

Each chapter (mundakas) is split into two sections (khandas) and this structure allows for a thorough exploration of spiritual concepts and the distinction between different types of knowledge. Through dialogues between a teacher and disciple, the text addresses the nature of reality and the means to attain spiritual enlightenment.

The Upanishad makes a clear distinction between apara-vidya (lower knowledge) and para-vidya (higher knowledge). The former encompasses the study of the Vedas, rituals, and worldly sciences, while higher knowledge pertains to the direct realization of Brahman, the ultimate reality. The path to moksha is the central theme of the Mundaka Upanishad, in which it emphasizes that true liberation is achieved through para-vidya – self-realization and understanding one's unity with Brahman.

While it highlights the importance of renunciation, meditation, and devotion in the pursuit of higher knowledge, it also encourages seekers to go beyond ritualistic practices and focus on inner transformation.

Here are some example texts from the Muṇḍaka Upanishad, along with explanations:

Muṇḍaka Upanishad 1.1.5:

द्वे विद्ये वेदितव्ये इति ह स्म यद्ब्रह्मविदो वदन्ति परा चैवापरा च।

Dve vidye veditavye iti ha sma yadbrahmavido vadanti parā caivāparā ca.

Translation:

"There are two types of knowledge to be known — lower and higher."

Explanation:

This verse introduces the concept of two types of knowledge — lower (apara) and higher (para). It suggests that spiritual seekers should strive to understand both forms of knowledge, indicating the importance of transcending worldly knowledge to attain higher spiritual wisdom.

Muṇḍaka Upanishad 2.2.8:

स यो ह वै तत्परं ब्रह्म वेद ब्रह्मैव भवति।

Sa yo ha vai tatparam brahma veda brahmaiva bhavati.

Translation:

"He who knows that supreme Brahman verily becomes Brahman."

Explanation:

This verse asserts the transformative power of realizing the supreme Brahman. It suggests that one who attains knowledge of Brahman transcends individual identity and merges with the ultimate reality. This realization leads to the attainment of liberation (moksha).

Muṇḍaka Upanishad 3.2.9:

सर्वं कर्माखिलं पार्थ ज्ञाने परिसमाप्यते।

Sarvam karmaakhilam pārtha jñāne parisamāpyate.

Translation:

"All actions are enveloped in knowledge, O Partha (Arjuna)."

Explanation:

This verse encapsulates the teaching that true knowledge encompasses and transcends all actions. It implies that spiritual wisdom leads to the understanding that all actions ultimately occur within the realm of knowledge. By realizing this, one attains liberation from the bondage of karma (action) and achieves spiritual fulfillment.

The Upanishads' philosophical depth, spiritual insights, and practical guidance make it a timeless and essential text in the Vedic tradition. Such deep philosophical inquiry into the nature of existence, the cosmos, and the ultimate truth are foundational to Vedantic philosophy and continue to inspire spiritual seekers.

6. Māṇḍūkya - Atharva Veda

The Mandukya Upanishad is an essential philosophical text found within the Atharva Veda. It is dedicated to the analysis of the syllable "Om" (A, U, M) and their symbolic representation of different states of consciousness and the ultimate reality.

Short but profound, the Mandukya Upanishad is the shortest of the major Upanishads, consisting of just 12 verses. Despite its brevity, it offers deep insights into the nature of reality and consciousness and is a key text for Advaita Vedanta, a non-dualistic school of thought.

It teaches that the individual self (Atman) and the ultimate reality (Brahman) are one and the same, transcending all dualities. The Upanishad describes four states of consciousness: waking (Jagrat), dreaming (Swapna), deep sleep (Sushupti), and the fourth state (Turiya), which is pure consciousness. It also provides profound philosophical insights in a concise form, making it a central text for those studying and practicing Vedanta.

Here are some example texts from the Māṇḍūkya Upanishad, along with explanations:

Māṇḍūkya Upanishad 1.1:

ओंकार एव चोदन्धर्मः ।

Oṁkāra eva codandharmaḥ.

Translation:

"The syllable Om is the supreme Brahman."

Explanation:

This opening verse establishes the significance of the syllable "Om" as the ultimate reality, equating it with Brahman, the supreme cosmic consciousness. It suggests that the entire universe manifests from and dissolves into the primal sound of Om.

Māṇḍūkya Upanishad 1.7:

नान्तःप्रज्ञं न बहिष्प्रज्ञं नोभयतःप्रज्ञं न प्रज्ञानघनं न प्रज्ञं नाप्रज्ञम् ।

अदृष्टमव्यवहार्यमग्राह्यमलक्षणमचन्ित्यमव्यपदेश्यं एकात्मप्रत्ययसारं प्रपञ्चोपशमं शान्तं शविमद्वैतं चतुर्थं मन्यन्ते स आत्मा स वज्िञेयः ॥

Nāntaḥprajñaṁ na bahiṣprajñaṁ nobhayataḥprajñaṁ na prajñānaghanaṁ na prajñaṁ nāprajñam ıṛṣṭamavya-vahāryamagrāhyamalakṣaṇamachintyamavyapadeśyam ekātmapratyayasāraṁ prapañcopaśamaṁ śāntaṁ śivamadvaitaṁ caturthaṁ manyante sa ātmā sa vijñeyaḥ ॥

Translation:

"It is neither the gross, nor the subtle, neither the conscious, nor the unconscious. It is unperceivable, ineffable, unthinkable, and ungraspable. It is the essence of the individual self, the cessation of all phenomena, serene, auspicious, and non-dual. This is to be known as the Self, and this is to be realized."

Explanation:

This verse describes the nature of the ultimate reality, which transcends all dualities and conceptualizations. It is beyond the realm of perception and cognition, existing as the essence of the individual self and the substratum of the universe. The verse emphasizes the importance of realizing this non-dual reality as the true Self.

Māṇḍūkya Upanishad 1.8:

सर्वं ह्येतद् ब्रह्मायमात्मा ब्रह्म सोऽयमात्मा चतुष्पात्।

Sarvaṁ hyetad brahmāyamātmā brahma so'yamātmā chatuṣpāt ।

Translation:

> *"All this is Brahman. This Self (Atman) is Brahman. This same Self has four quarters."*
>
> Explanation:
>
> This verse asserts the identity of Brahman with the entire universe and the individual self. It states that the ultimate reality pervades everything and that the individual self (Atman) is none other than Brahman itself. Additionally, it introduces the concept of the four states of consciousness: waking, dreaming, deep sleep, and the transcendental state (Turiya).

This Upanishads' concise yet deep teachings make it a foundational text in Advaita Vedanta and an essential guide for those pursuing spiritual knowledge and self-realization.

7. Taittirīya - Yajur Veda

The Taittiriya Upanishad is a significant philosophical text found within the Taittiriya branch of the Yajur Veda. It is divided into three chapters (shakhas) known as Shiksha Valli, Brahmananda Valli, and Bhrigu Valli, each containing rich teachings on various aspects of spirituality, meditation, and the nature of reality.

The Shiksha Valli concentrates on phonetics, pronunciation, and the importance of Vedic recitation, providing foundational knowledge for understanding and chanting the Vedas correctly. Brahmananda Valli explores the nature of Brahman (the ultimate reality) and the bliss associated with realizing this supreme truth. It discusses the layers of human existence and the path to spiritual enlightenment. Bhrigu Valli presents a dialogue between the sage Bhrigu and his father Varuna, focusing on the experiential understanding of Brahman through the practice of tapas (austerity) and meditation.

The Taittiriya Upanishad offers a holistic approach to spirituality, addressing the theoretical, practical, and experiential aspects of the spiritual path. The text introduces the concept of the five sheaths (koshas) that cover the true self (Atman): the physical body (Annamaya Kosha), the vital breath (Pranamaya Kosha), the mind (Manomaya Kosha), the intellect (Vijnanamaya Kosha), and bliss (Anandamaya Kosha). Understanding these layers is crucial for self-realization.

It also elaborates on the nature of Brahman and the bliss that comes from its realization, emphasizing that true happiness and fulfillment lie in understanding and experiencing the ultimate reality. There are plenty of practical instructions on meditation, ethical living, and the importance of self-discipline, guiding seekers on how to incorporate spiritual practices into their daily lives.

Here are some example texts from the Taittiriya Upanishad, along with explanations:

Taittirīya Upanishad 2.1.1:

आत्मावारे द्राधं नित्यं ।

Ātmāvāre drādhaṁ nityam ।

Translation:

"Hold onto the self steadfastly."

Explanation:

This verse emphasizes the importance of steadfastly holding onto the knowledge of the self (Atman). It encourages seekers to cultivate a continuous awareness of their true nature, recognizing the self as eternal and unchanging amidst the transient nature of the world.

Taittirīya Upanishad 2.4.1:

तमेव विदित्वाऽतिमृत्युमेति नान्यः पन्था विद्यतेऽयनाय ।

Tameva viditvā'timṛtyumeti nānyaḥ panthā vidyate'yānāya ।

Translation:

"Knowing Him alone, one transcends death. There is no other path to liberation."

Explanation:

This verse declares the knowledge of the supreme reality (Brahman) as the ultimate means of transcending the cycle of birth and death. It emphasizes that liberation can be attained only through realizing the true nature of Brahman, highlighting the exclusivity of this path to spiritual liberation.

Taittirīya Upanishad 3.1.1:

सत्यं ज्ञानं अनन्तं ब्रह्म ।

Satyam jñānam anantam brahma.

Translation:

"Truth, knowledge, and infinity is Brahman."

Explanation:

This concise statement encapsulates the nature of Brahman, the ultimate reality. It defines Brahman as Truth (Satyam), Knowledge (Jnanam), and Infinity (Anantam), highlighting its essential attributes as the substratum of existence, the source of wisdom, and the infinite, unbounded reality beyond all limitations.

Through its three chapters, the Taittiriya Upanishad provides profound teachings on the layers of human existence, the nature of Brahman, and practical guidance for spiritual practice. Its blend of theoretical knowledge, practical instructions, and experiential insights makes it a timeless and essential text in the Vedic tradition.

8. Aitareya - Rig Veda

The Aitareya Upanishad is a significant philosophical text found within the Aitareya Shakha (branch) of the Rig Veda. It contains cosmological teachings and describes the creation of the universe from the cosmic being (Purusha). Purusha is not only the cause but also the substance of the universe, embodying both immanence and transcendence. It also discusses the nature of the self and its relationship to the ultimate reality (Brahman), emphasizing that the true self resides as the eternal, unchanging essence within all beings.

The Aitareya Upanishad contains profound philosophical insight and its teachings are foundational to Vedanta philosophy, as well as significantly influencing subsequent spiritual and philosophical thought in India. It also provides a framework for spiritual practice, emphasizing meditation and self-inquiry as means to attain self-realization and liberation.

Here are some example texts from the Taittirīya Upanishad, along with explanations:

Aitareya Upanishad 1.1.1:

आत्मा वा इदमेक एवाग्र आसीन्नान्यत्किंचन मषित्।

Ātmā vā idameka evāgra āsīn nānyat kiṃcana miṣat.

Translation:

"In the beginning, this universe was the Self alone. There was nothing else whatsoever that winked."

Explanation:

This verse expounds the concept of the universe emanating from the primordial Self (Atman). It asserts that initially, before creation, only the Self existed, and there was nothing apart from it. This highlights the non-dualistic nature of reality, where the Self is the ultimate substratum of all existence.

Aitareya Upanishad 2.3:

स तपोऽतप्यत।

Sa tapo'tapyata.

Translation:

"He performed austerities."

Explanation:

This verse describes the process of creation wherein the Supreme Being (Brahman) engaged in tapas (austerities) before manifesting the universe. Tapas represents the intense spiritual effort or penance undertaken by the creator to bring forth the cosmos, underscoring the active role of Brahman in the creative process.

Aitareya Upanishad 3.1:

यो वै स तत्परमात्मा तत्त्वमसि श्वेतकेतो।

Yo vai sa tatparamatma tat tvamasi shvetaketo.

Translation:

"He who is the Supreme Self, that thou art, O Shvetaketu."

Explanation:

This famous statement encapsulates the essence of Advaita Vedanta philosophy, asserting the identity of the individual self (Atman) with the ultimate reality (Brahman). It emphasizes the realization that the true nature of the self is identical to the supreme consciousness, represented by the Mahavakya (great dictum) "Tat Tvam Asi" (That Thou Art).

Despite its ancient origins, the Aitareya Upanishad's exploration of consciousness and existence remains relevant. Its teachings transcend cultural and temporal boundaries, offering insights applicable to modern seekers. The text integrates cosmology, psychology, and spirituality, presenting a holistic view of existence that reflects the Vedic approach to knowledge, where different domains of understanding are interconnected.

9. Chāndogya - Sama Veda

The Chandogya Upanishad is one of the longest and most important Upanishads found within the Chandogya branch of the Samaveda. It consists of eight chapters (prapathakas) and contains a wide range of philosophical teachings and spiritual practices, covering topics such as the nature of Brahman, the importance of meditation, and the significance of rituals.

The philosophical teachings found within the text extensively explore the concept of Brahman, the ultimate reality, describing Brahman as the source and essence of all existence, transcending all dualities and distinctions. The Upanishad also emphasizes the practice of meditation (dhyana) as a means to realize Brahman, outlining various meditative techniques and contemplations on sacred syllables like "Om" to aid in spiritual awakening.

While the text acknowledges the importance of rituals, it often interprets them symbolically, encouraging practitioners to seek deeper, inner meanings rather than merely perform external rites. The dialogues and stories within the text serve as inspirational guides, illustrating the path of inquiry, devotion, and realization through dedication to spiritual growth.

Here are some example texts from the Chandogya Upanishad, along with explanations:

> Chāndogya Upanishad 6.8.7:
>
> तत्त्वमसि।
>
> Tat tvamasi.
>
> Translation:
>
> *"That thou art."*
>
> Explanation:
>
> This concise statement, known as the Mahavakya (great dictum), encapsulates the essence of Advaita Vedanta philosophy. It asserts the identity of the individual self (Atman) with the ultimate reality

(Brahman), emphasizing the non-dualistic nature of existence and the unity of all beings.

Chāndogya Upanishad 1.1.1:

सत्‌यं ज्‌ञानं अनन्‌तं ब्‌रह्‌म।

Satyam jñānam anantam brahma.

Translation:

"Brahman is Truth, Knowledge, and Infinity."

Explanation:

This verse defines Brahman, the ultimate reality, as Satyam (Truth), Jnanam (Knowledge), and Anantam (Infinity). It highlights the essential attributes of Brahman, emphasizing its role as the substratum of existence, the source of wisdom, and the infinite, unbounded reality beyond all limitations.

Chāndogya Upanishad 7.24.1:

तदेतत्‌सत्‌यं स आत्‌मा तत्‌त्‌वमसि श्‌वेतकेतो।

Tad etat satyam sa ātmā tat tvam asi, śvetaketo.

Translation:

"That is the truth, that is the Self (Atman), thou art That, O Shvetaketu."

Explanation:

This verse reinforces the teaching of identity between the individual self (Atman) and the ultimate reality (Brahman). It instructs Shvetaketu to realize his true nature as identical to Brahman, affirming the non dualistic nature of the self and the universe.

The Chandogya Upanishad offers rich and varied teachings on the nature of Brahman, the practice of meditation, and the symbolic significance of rituals. Its comprehensive and profound insights continue to inspire and guide spiritual seekers, contributing significantly to the philosophical foundation of Vedanta and the broader spiritual landscape of India.

10. Bṛhadāraṇyaka - Yajur Veda

Last but not least, the Brihadaranyaka Upanishad is one of the oldest and most voluminous Upanishads found within the Brihadaranyaka branch of the Yajur Veda. It consists of six chapters, each one containing profound teachings on the nature of reality, the self, and the ultimate truth (Brahman). It also includes philosophical dialogues and discussions between sages and seekers.

The Upanishads' profound teachings deal with the nature of reality, Neti Neti (Not This, Not That), and the nature of the self, highlighting that true knowledge and enlightenment come from understanding the self's oneness with Brahman. As one of the oldest Upanishads, the Brihadaranyaka Upanishad holds a crucial place in the development of Indian philosophical and spiritual thought, with its teachings influencing various schools of philosophy, particularly Vedanta.

Here are some example texts from the Brihadaranyaka Upanishad, along with explanations:

Bṛhadāraṇyaka Upanishad 1.3.28:

तत्त्वमसि, श्वेतकेतो ।

Tat tvam asi, śvetaketo.

Translation:

"That thou art, O Shvetaketu."

Explanation:

This verse is part of a dialogue between Sage Uddalaka and his son Shvetaketu, where the father imparts profound spiritual wisdom to his son. It emphasizes the identity of Shvetaketu's individual self (Atman) with the ultimate reality (Brahman). Through this teaching, Shvetaketu is guided to realize his true nature and transcend the limitations of the ego, leading to spiritual enlightenment.

Bṛhadāraṇyaka Upanishad 2.1.20:

अहं ब्रह्मास्मि।

Aham brahmāsmi.

Translation:

"I am Brahman."

Explanation:

This declaration represents a profound realization of the true nature of the self. It asserts the identity of the individual self (Atman) with Brahman, the ultimate reality. By recognizing oneself as Brahman, one transcends the limitations of the ego and attains liberation (moksha).

Bṛhadāraṇyaka Upanishad 2.4.6:

असतो मा सद्गमय। तमसो मा। ज्योतिर्गमय। मृत्योर्मा अमृतं गमय।

Asato mā sadgamaya. Tamaso mā jyotirgamaya. Mṛtyormā amṛtaṁ gamaya.

Translation:

"Lead me from the unreal to the real, from darkness to light, from death to immortality."

Explanation:

This famous prayer embodies the seeker's aspiration for spiritual realization and liberation. It expresses the desire to transcend the ignorance and limitations of worldly existence (asat) and attain the ultimate truth (sat). It seeks guidance from darkness (ignorance) to light (knowledge) and from mortality to immortality, symbolizing the journey from bondage to liberation.

The Brihadaranyaka Upanishad is a foundational text within the Vedic tradition, offering profound insights into the nature of reality, the self, and ultimate truth. Its rich philosophical dialogues and comprehensive teachings continue to inspire and guide spiritual seekers, making it one of the most important works in Indian spiritual literature.

These ten principal Upanishads are revered for their depth of wisdom, spiritual insights, and profound teachings on the nature of existence and the path to self-realization. They continue to be studied and revered by seekers of truth and wisdom in the Hindu tradition and beyond.

In the next few chapters, we will further elaborate on the key concepts and constructs from within the Upanishads that make up the core of the philosophical construct of Vedanta.

Illumination or Hallucination: Does it really matter?

About two decades ago, I had an opportunity to study with a breatharian. After I had spent quite a while with him learning the science of breath, he suddenly suggested to me one day that I should take a hike along a particular route at the foothills of the upper Himalayas. There, I would come across a sacred temple, where I should stay for a couple of days before returning to him to share what I had learnt. He would then begin the next phase of my training.

I started out on that journey early the next day, spending almost seven hours hiking down to a village where I planned to spend the night before beginning the journey further. This meant I would need to follow a river by walking along its banks.

That journey along the river bank proved to be a lot harder than I expected. Very soon my feet were blistered by walking the stony terrain as I was only wearing regular shoes, not hiking boots. Just as I was beginning to rearrange the timeline in my mind for that trip, it suddenly started pouring with rain, as if the heavens had opened up. Anticipating this, I had brought with me a raincoat and quickly opened my bag to put the raincoat on before continuing my walk. After a while, it somehow seemed that my backpack felt lighter and, after a few more paces, I stopped to check. I realized I had not completely closed it and, somewhere along the journey, my food packet (some *puris* – an Indian bread and some nuts) had fallen out. This worried me a bit as I knew the next village was at least a 5 hour hike away in this weather and there was a good chance I might not make it before nightfall.

However, I kept walking, coming up to a bridge I was supposed to cross to start the climb down to the village. But to my shock, the makeshift bridge had been swept away and only some planks were left dangling across the fast-moving river that had swelled up after the rain. I didn't know how to swim back then and I still don't. This meant I was stuck – I was 6 hours from where I started, and 5 hours from where the next stop was. The journey had doubled due to the need to walk slowly through the rain and I had no food, with only a little water.

Anyone who knows me can vouch that my jungle survival skills are close to zero. Actually, my skills to survive even a camping trip are non-existent and now this was a real issue. Night was coming in fast so I decided to try to find some shelter from the rain under a tree and think about what to do next. It must have only been 5 minutes after I sat down that I realized how tired I was and I dozed off, waking up almost an hour later to a very dark outdoors. The clouds and early sunset had turned the evening sky pitch black. I did have a torch and flashlight-equipped helmet so decided to look around. About 20 minutes later, it became apparent that there was no chance of going back or walking on further tonight, which meant I had to spend the night where I was. The rain had stopped, it wasn't cold, and the sun would probably come up in a few hours, so I decided to wait it out. The only thing bothering me was my growling empty stomach.

After a while, I saw a very bright light flickering some distance uphill. It was unmistakably the light from a fire and I decided to walk up to it. A brisk twenty minutes walk later, I arrived at a cave where two Sadhus were sitting inside, meditating. Almost as if they were expecting me, one of them opened his eyes the moment I arrived and gestured to me to sit on a grass cushion next to the fire. Within the next few minutes, I was dry and warm. The Sadhus, who by this time had opened their eyes, then asked me, "Are you ready?" I almost wanted to ask, "Ready for what?" but found myself saying yes.

I was then asked to take some water from the Kammandal, (a container that Sadhus carry) and drink. It must have only been two or three sips but I found myself feeling a little dizzy so I decided to lie down. Right in front of my eyes, I saw the scene of the cave change. There were now four sadhus instead of two, and another ten pupils. It was a class in which sadhus were teaching the pupils about how the Vedas were compiled and why now the Upanishads are being written to better explain the philosophical construct of the Vedas. The conversation was in a version of Sanskrit I did not recognize but understood quite clearly. I listened intently for about an hour and a half in amazed silence, absorbing everything. After a time, one of the sadhus turned to me and said that it was enough for me and that I should drink some more water. I followed the instructions and took some more sips. Just as I did so and placed the earthen cup down next to the Kammandal, a flash made me sit up alertly... I had woken up to the bright sun shining in my face.

I was back at the bank of the river where I had sat down to rest. The morning sun had fully risen and the sky was clear. I felt quite disoriented and somewhat of a state of shock, not knowing what had just happened. I decided to hike back to the village where I had spent the night before, arriving there by noon. To my utter surprise, I found my breatharian teacher waiting for me at the temple, right at the entrance of the village.

He smiled and asked me, *"How was it?"*

"I don't know," I replied.

"When you are ready, you will know and remember. For now, just rest," he said, and asked me to stay in the village for a couple of days before coming back to his Kuti higher up.

When I eventually went back to him a couple of days later, he didn't mention anything about my experience except for these two lines: *Not all temples are physical. Some are doors that open for you when you are ready.*

So what had happened to me? Did I dream it all? Was I delusional? Did I have a deep spiritual experience of time travel or astral visit to a past period?

I still don't know, and I realize it doesn't matter. Since that day, though, the Upanishads literally speak to me, often surprising me with what I know, beyond what is written – though I have no clue why or how I know. And for that, I am supremely grateful to Baba ji!

"The patterns of the universe are also the patterns of our own consciousness. The Upanishads are the ancient sages' attempt to explore and understand the nature of these patterns."

– Fritjof Capra

PART I:

UNVEILING THE COSMIC REALITY

One of the quotes from Upanishads I find most profound is one that exemplifies the notion of cosmic reality. It is from the Chandogya Upanishad and reads:

"Verily, my dear, in the beginning this universe was Being alone, one only without a second." *(Chapter 6, Section 2, Verse 1)*

> "प्रवृत्तिं च निवृत्तिं च जना न विदुरासुराः ।
>
> न शौचं नापि चाचारो न सत्यं तेषु विद्यते ॥»
>
> "Pravṛittiṁ cha nivṛittiṁ cha janā na vidurāsurāḥ
>
> Na śauchaṁ nāpi chāchāro na satyaṁ teṣu vidyate."

This verse is often cited as a fundamental expression of the non-dualistic perspective in Vedanta philosophy, emphasizing the ultimate unity of existence, and of course the new age euphemism of "*All are one.*" The notion of cosmic reality in Vedanta suggests that beneath the manifold appearances of the universe lies an underlying unity, represented by the concept of Brahman.

Brahman is described as the unchanging, infinite, and eternal essence that serves as the substratum of the entire universe. It transcends all limitations, including time, space, and causality, and is beyond all attributes, forms, and conceptualizations. Think of Brahman as an ocean in which we are all like waves: distinct and separate on the surface but ultimately one with the ocean itself.

Imagine that there is a fundamental unity that underlies the apparent diversity of existence. While the world around you may seem fragmented and divided into various forms, entities, and experiences, Vedanta explains that all of these are mere manifestations of the one underlying reality of Brahman.

From the smallest atom to the vast cosmos, everything in creation is sustained by and ultimately merges back into Brahman.

This understanding of cosmic reality has profound implications for how you might perceive and relate to the world. Rather than seeing the universe as a collection of separate and isolated entities, Vedanta invites you to recognize the interconnectedness and unity that pervades all of existence. This recognition of unity leads to a shift in perspective, fostering feelings of interconnectedness, empathy, and reverence for all life.

The concept of cosmic reality in Vedanta also suggests that the ultimate purpose of our existence is to acknowledge this underlying unity with Brahman. Through spiritual practice, self-inquiry, and contemplation, we can seek to transcend the limitations of the ego and manifest our essential unity with the ultimate reality.

This awakening is considered the highest goal of human life and the key to attaining spiritual liberation (moksha) in Vedanta philosophy.

"Vedanta teaches that consciousness is singular, all happenings are played out in one universal consciousness and there is no multiplicity of selves."

– Erwin Schrödinger

CHAPTER 4

BRAHMAN: UNSEEN, UNBOUNDED, UNCONSTRAINED

In Vedanta, the concept of Brahman is more than just a philosophical abstraction; it's a profound invitation to explore the deepest mysteries of existence. In a journey that defies time and place ,it allows us to go beyond what we understand by reality and takes us on an illuminating quest.

Imagine Brahman as the cosmic heartbeat, the timeless rhythm that pulses through every atom, every thought, and every breath of the universe. It's the unchanging, boundless essence from which all creation emerges, sustains itself, and ultimately returns. Brahman is the cosmic architect, designing the structures of reality with blueprints of infinite potentiality, transcending the confines of time, space, and causality. It's the silent witness to the dance of the cosmos, the unseen force guiding the symphony of life.

Contemplating Brahman invites us to transcend the limitations of our ordinary perception and delve into the infinite depths of our own consciousness. It's an invitation to recognize the interconnectedness of all beings, to see beyond the illusions of separateness and duality, and to awaken to the profound unity that underlies the diversity of existence.

Through realizing Brahman, we come to understand that we are not merely finite beings bound by the constraints of our individual identities, but infinite expressions of the same divine essence that animates the entire cosmos.

Embracing the concept of Brahman is to embark on a journey of self-discovery and spiritual awakening, one that leads us beyond the confines of our egoic mind and into the boundless expanse of our true nature. It's a journey of liberation, of realizing our inherent divinity and reconnecting with the source of all existence.

In the light of Brahman, every moment becomes a sacred opportunity for communion with the divine, every experience a gateway to profound wisdom and bliss.

Despite belonging to one of the most ancient traditions of wisdom mankind has ever known, the idea of Brahman continues to hold so much significance in the context of contemporary life.

Its timeless allure resonates deeply with the challenges and aspirations of modern humanity. In a world that often feels fragmented and disconnected, where technology can sometimes distance us from our own humanity and the natural world around us, the concept of Brahman offers a profound antidote — an invitation to reconnect with the deeper currents of existence.

Living in a society driven by materialism and external success, the concept of Brahman reminds us of the futility of chasing after transient pleasures and possessions. It invites us to shift our focus from the external to the internal, from the pursuit of fleeting happiness to the cultivation of lasting inner peace and fulfillment. We all know that neither wealth nor instant gratification alone can bring us happiness. The incessant desire to experience the next big thing soon becomes jaded and, very often, we are left feeling empty. It's this void that the Vedantic concept of Brahman can fill, reminding us of the importance of slowing down, of pausing to reflect and reconnect with the essence of who we truly are.

This world has seen, and continues to witness, its share of conflicts and tensions, with polarization and exclusion evident in every walk of life. The idea of Brahman offers a powerful vision of unity and interconnectedness. It transcends the boundaries of race, religion, and nationality, reminding us that at the deepest level, we are all expressions of the same divine essence. Embracing this perspective can inspire us to cultivate compassion, empathy, and understanding toward others, fostering a sense of global citizenship and collective responsibility for the well-being of all.

Moreover, in an age of uncertainty and rapid change, where many of us feel adrift in a sea of information overload and societal upheaval, the concept of Brahman provides a stable anchor — a timeless refuge amidst the storms of life. It offers a perspective that transcends the transient fluctuations of the external world, inviting us to find solace and strength in the unchanging essence that lies at the core of our being.

Ultimately, what makes the idea of Brahman so appealing in contemporary life is its timeless relevance and universal resonance. It speaks to the deepest longings of the human heart — the yearning for meaning, connection, and transcendence — and offers a path towards greater harmony, wisdom, and fulfillment in our individual lives and the world at large.

But what exactly do the Upanishads say?

Much of what we read about the commentaries or listen to in lectures about Vedanta is perhaps an interpretation of other interpretations of what has been said across different Upanishads, carefully curated and put together by the speaker/author. I guess the researcher-scholar in me is a stickler about going back to the source and drawing wherever possible my own insights. Hence, I am quoting here for you directly from the Upanishads, with a very brief explanation, inviting you to contemplate the rest.

While the translation itself is an interpretation, it is my attempt to trigger your inquiry and imagination, rather than giving you only my own version. So, here is a closer look at different aspects of Brahman as *"Unseen, Unbounded, Unconstrained"*, as originally mentioned in the source texts.

Unseen:

Brahman is described as "unseen" because it cannot be perceived by the senses or grasped by ordinary modes of perception or cognition. It transcends all empirical categories and is beyond the realm of empirical observation. Brahman is not an object of perception but is the very substratum of all perception, existing beyond the limitations of the sensory world.

How does Vedanta talk about this idea of Brahman being unseen?

We only have to look at the Upanishads to find out. Here are some noteworthy quotes about the nature of Brahman:

Unseen Nature of Brahman:

- "That which cannot be seen, yet through which one sees, that is Brahman." (Mundaka Upanishad 1.1.7)

  This quote encapsulates the essence of Brahman as the ultimate reality that transcends the limitations of perception. Brahman is described as invisible, beyond the grasp of the senses, yet it is the very source of our ability to perceive and experience the world. It suggests that Brahman is not an object to be perceived externally but is the very essence of our consciousness, the inner light that illuminates all experience. In other words, Brahman is the underlying reality that enables us to see and comprehend the world around us, even though it cannot be perceived as an object itself.

- "The Self (Atman) is indeed Brahman, but through ignorance people do not know It, thinking 'I know Brahman.'" (Brihadaranyaka Upanishad 1.4.10)

  This quote highlights the fundamental unity between the individual self (Atman) and the universal reality (Brahman). It suggests that Brahman is not something external to ourselves but is our true nature, the innermost essence of our being. However, due to ignorance and delusion, people fail to recognize this inherent unity and mistakenly identify themselves with their limited egoic self. They believe they know Brahman as something separate from themselves, failing to realize that Brahman is their very essence. This quote underscores the importance of transcending ignorance and realizing the true nature of the self as identical with Brahman, leading to liberation and spiritual enlightenment.

Beyond Ordinary Perception:

- "This Self cannot be attained through study of the Vedas, nor through intelligence, nor through much learning. He who chooses the Self attains It; to him verily It reveals Its true nature." (Katha Upanishad 1.2.23)

  This quote emphasizes the idea that true realization of the Self (Atman) cannot be achieved solely through external knowledge or intellectual pursuits. Despite the importance of studying sacred texts like the Vedas or cultivating intelligence, the ultimate realization of the Self transcends these conventional means. Instead, it requires a deep inner commitment and personal choice to seek and align oneself with the Self. The Self, being the essence of one's being, reveals its true nature to those who earnestly seek it with sincerity and dedication. In essence, the path to Self-realization involves a profound inner journey of self-discovery and spiritual awakening, guided by the sincere desire to know one's true nature.

- "The wise man, having realized the Self as indestructible, eternal, unborn, and immutable, and knowing Him as the one who transcends all material experiences, grieves no more." (Katha Upanishad 2.22)

This quote speaks to the profound peace and liberation that comes from realizing the true nature of the Self. It describes the state of the wise individual who has transcended the illusion of the ego and recognized the eternal, unchanging essence within. By understanding the Self as indestructible, beyond birth and death, and transcending all material experiences, the wise person is freed from the cycle of suffering and sorrow. Such a realization brings about a deep sense of contentment, as the wise person no longer identifies with transient phenomena or worldly concerns. Instead, they abide in the awareness of their eternal nature, untouched by the fluctuations of the material world. This passage underscores the transformative power of Self-realization, which leads to a state of profound peace, inner stability, and freedom from suffering.

Transcendence of Empirical Categories:

- "It is not comprehended by the eye, nor by speech, nor by the other senses, nor by penance or good works. A man becomes pure through serenity of intellect and meditation." (Mundaka Upanishad 3.1.8)

  This quote highlights the ineffable nature of Brahman, the ultimate reality. It suggests that Brahman cannot be fully grasped or understood through the ordinary means of perception, such as sight, speech, or the other senses. Additionally, it cannot be attained through external rituals or acts of merit alone. Instead, true purity and understanding come through the cultivation of inner tranquility, clarity of intellect, and deep contemplation through meditation. This emphasizes the importance of inner spiritual practices in gaining insight into the nature of Brahman, transcending the limitations of the physical senses, and reaching a deeper level of awareness.

- "Those who realize Brahman know that all the senses are rooted in the mind and that the mind is rooted in Brahman." (Katha Upanishad 2.3.10)

  This quote elucidates the relationship between Brahman, the mind, and the senses. It suggests that those who attain realization of Brahman understand that the mind is the intermediary between the senses and the ultimate reality. The senses, which perceive the external world, are rooted in the mind, which interprets and processes sensory information. However, beyond the mind lies Brahman, the source and foundation of all existence. Realizing Brahman involves recognizing that the mind, with its thoughts and perceptions, ultimately arises from and is sustained by Brahman. Therefore, by transcending the mind and attaining union with Brahman, one can attain true insight and liberation from the illusions of the material world. This quote underscores the importance of understanding the relationship between the mind, the senses, and Brahman in the quest for spiritual realization and enlightenment.

The Substratum of Perception:

- "The self-existent Lord pierced the senses to turn outward. Thus we look to the world outside and see not the Self within us. A sage withdrew his senses from the world of the senses and, with his mind, turned inwards and beheld the glory of the Self." (Katha Upanishad 2.1.1)

  This quote from the Katha Upanishad beautifully captures the essence of the human experience and the journey toward self-realization. It begins by describing how the "self-existent Lord," often interpreted as the divine or the universal consciousness, has endowed human beings with senses that naturally orient outward, towards the external world. This outward orientation leads us to engage with the world around us, becoming absorbed in its myriad experiences and distractions. In doing so, we become disconnected from the inner reality of the Self, which remains obscured by the constant flux of sensory perceptions.

  However, the quote offers a profound insight into the transformative power of wisdom and self-awareness. It describes how a sage, through wisdom and spiritual practice, consciously withdraws their senses from the external world. By turning their attention inward, the sage redirects their focus toward the inner realm of consciousness, seeking to behold the true essence of the Self that resides within.

In this inward journey, the sage transcends the limitations of the external senses and the distractions of the material world. Through the power of introspection and meditation, they are able to pierce through the veil of illusion and discover the inherent glory of the Self — the eternal, unchanging essence that lies beyond the fluctuations of the external reality.

This quote serves as a profound reminder of the importance of introspection and spiritual practice in the pursuit of self-realization. It teaches us that true fulfillment and liberation come not from external pursuits and sensory gratification, but from turning inward and awakening to the timeless truth of our own being.

The above quotes illustrate the profound nature of Brahman as beyond ordinary perception, transcending empirical categories, and yet being the substratum of all perception. They stress the importance of self-realization and inner wisdom in understanding Brahman, which cannot be attained through mere intellectual study or sensory perception but through direct spiritual experience and awakening of the true Self (Atman).

Unbounded:

Brahman is "unbounded" because it is limitless, infinite, and without boundaries. It is not confined to any particular form, shape, or dimension but encompasses everything in existence. Brahman is beyond all distinctions and limitations, including those of time, space, and causality. It pervades everything in creation and transcends all dualities and opposites.

Here are some references to the unbounded nature of Brahman, as mentioned in the Upanishads:

Limitless and Infinite Nature:

- "Brahman is infinite, and this universe is infinite. The infinite proceeds from the infinite. If you subtract the infinite from the infinite, the infinite remains alone." (Taittiriya Upanishad 2.1.1)

  Here, we find an illustration of the concept of Brahman as infinite and omnipresent. It suggests that Brahman, the ultimate reality, is not limited by space or time and encompasses all of existence. The universe itself is an expression of this infinite nature of Brahman. Even if one were to attempt to subtract or diminish the infinite from the infinite, Brahman would remain unchanged and unaffected, as it is beyond any form of reduction or limitation. This quote underscores the incomprehensible vastness and boundless nature of Brahman, which transcends human understanding.

- "Brahman is bliss. He is supreme bliss. From bliss all beings originate, by bliss they are sustained, and to bliss they return." (Taittiriya Upanishad 3.6.1)

In this quote, Brahman is presented not only as infinite but also as the source of supreme bliss. It suggests that Brahman is the ultimate source of happiness and joy, from which all beings originate and to which they ultimately return. This idea reflects the Upanishadic concept of Ananda, or bliss, as an inherent quality of Brahman. It implies that the true essence of existence is one of profound happiness and contentment, and that all beings are sustained by this blissful nature of Brahman. Furthermore, it suggests that the ultimate goal of life is to realize this inherent bliss within oneself and to return to it through spiritual realization and enlightenment.

Absence of Boundaries:

- "The great, unborn Self, undecaying, undying, immortal, fearless, is indeed Brahman." (Brihadaranyaka Upanishad 4.4.25)

  This quote encapsulates the essence of Brahman as described in the Brihadaranyaka Upanishad. It highlights Brahman as the eternal, unchanging, and limitless reality that transcends birth and death. The Self, referred to here as unborn and undecaying, is synonymous with Brahman, signifying the divine essence within each individual. It emphasizes the inherent immortality and fearlessness of the Self/Brahman, which remains unaffected by the transient phenomena of the world.

- "Where one sees nothing else, hears nothing else, understands nothing else, that is the Infinite. Where one sees something else, hears something else, understands something else, that is the finite. The Infinite is immortal, the finite is mortal." (Chandogya Upanishad 7.24.1)

  Here, we read of the contrast between the Infinite (Brahman) with the finite (the material world). It suggests that true realization of the Infinite occurs when one transcends the limitations of sensory perception and intellectual understanding, attaining a state of pure awareness where nothing else is perceived or comprehended. In contrast, the finite world is characterized by the perception of diversity and separation, where objects and experiences are perceived as distinct entities. The quote emphasizes the immortality of the Infinite (Brahman) compared to the mortality of the finite, highlighting the eternal nature of the ultimate reality beyond the realm of material existence.

Encompassing Everything in Existence:

- "Brahman, though hidden in all beings, does not shine forth. It is seen, however, by those seers of subtle vision." (Mundaka Upanishad 3.1.2)

  This quote elucidates the concealed nature of Brahman within all beings. Despite its omnipresence, Brahman remains veiled from ordinary perception. However, those endowed with profound insight and intuitive wisdom, often referred to as seers of subtle vision, are capable of perceiving Brahman. This suggests that the realization of Brahman requires a deeper level of awareness and spiritual insight beyond the ordinary senses. It emphasizes the importance of developing inner clarity and intuition to behold the transcendent reality of Brahman hidden within the fabric of existence.

- "All this is Brahman. From It the universe comes forth, in It the universe merges, and in It the universe breathes." (Mundaka Upanishad 2.2.11)

  Here, Brahman is presented as the ultimate source, sustainer, and essence of the universe. It asserts that everything, including the entirety of existence, originates from Brahman, subsists within Brahman, and ultimately merges back into Brahman. This concept illustrates the interconnectedness and interdependence of all phenomena with the transcendent reality of Brahman. It suggests that Brahman is the underlying substratum that permeates every aspect of creation, from the tangible to the intangible. The passage underscores the omnipresence and omnipotence of Brahman as the cosmic principle that governs the unfolding and evolution of the universe.

Beyond Distinctions and Limitations:

- "As the one fire, after it has entered the world, though one, takes different forms according to whatever it burns, so does the internal Atman of all living beings, though one, takes a form according to whatever It enters, and exists also without." (Brihadaranyaka Upanishad 2.4.6)

This quote employs the metaphor of fire to illustrate the nature of the Atman (Self). Just as a single fire can manifest in various forms depending on what it burns, the Atman, though singular, takes on different individual forms in all living beings. It permeates each being and adapts to the specific circumstances and conditions it encounters, yet it remains unchanged and eternal. This example underscores the unity of the Atman across all beings and emphasizes its transcendental nature, existing beyond the realm of individual identities.

- "Brahman is devoid of attributes, and is unconditioned. The wise, freed from egoism and free from duality, beholds the glorious Atman." (Mundaka Upanishad 3.1.9)

  In the above, Brahman is described as attributeless and unconditioned, transcending all characteristics and limitations. It suggests that Brahman cannot be defined or confined by any particular qualities or attributes, as it is beyond the realm of dualistic concepts. Only the wise, who have transcended egoism and the perception of duality, are capable of truly beholding the glorious Atman — the innermost essence of the Self. This highlights the importance of spiritual liberation and self-realization in attaining a direct experience of Brahman, beyond the veil of individual identity and conceptual limitations.

Pervading Everything in Creation:

- "Brahman is all-pervasive. All beings exist in Brahman, but Brahman does not exist in them. Just as the all-pervasive space exists in a jar, but the jar does not exist in space." (Chandogya Upanishad 8.1.1)

  This quote vividly illustrates the omnipresence of Brahman, the ultimate reality. It emphasizes that Brahman encompasses all beings and phenomena within its boundless existence, yet remains transcendent and unaffected by them. The analogy of space within a jar elucidates this concept — just as space permeates the interior of a jar but is not confined or limited by it, Brahman permeates all of creation but is not constrained by it. This highlights the fundamental unity of existence within Brahman while emphasizing its transcendental nature beyond the limitations of individual forms.

- "He is all-pervading, radiant, bodiless, woundless, without muscles, pure and untouched by sin. He is the seer, all-knowing, transcendent, and self-existent. He has ordained the objects from beginning to end in their respective positions." (Svetasvatara Upanishad 3.16)

  Here again, Brahman is described as omnipresent and transcendent, possessing divine qualities beyond human comprehension. Brahman is depicted as pervading all of existence with radiant purity, devoid of physical attributes such as a body or wounds. Furthermore, Brahman is characterized as the ultimate seer and knower, transcending all limitations of knowledge and perception. It is the divine ordainer of the universe, orchestrating the arrangement and functioning of all objects and beings from the beginning to the end of time. This portrayal underscores the divine omnipotence and omniscience of Brahman, as well as its role as the foundational principle of existence.

These quotes from the Upanishads illustrate the limitless, infinite, and all-encompassing nature of Brahman, which transcends all boundaries, distinctions, and limitations. Brahman pervades everything in creation and yet remains beyond all dualities and opposites, serving as the ultimate substratum and source of the universe.

Unconstrained:

Brahman is "unconstrained" because it is free from any limitations, restrictions, or dependencies. It is self-existent, self-sufficient, and self-luminous, existing independently of anything else. Brahman is not subject to change, decay, or destruction but is the unchanging reality underlying all changes and transformations in the universe.

Here are some more examples from the sacred texts that describe the unconstrained nature of Brahman:

Self-Existent and Self-Sufficient Nature:

- "From Brahman all worlds proceed, in Brahman they are sustained, and unto Brahman they return. Brahman is the ultimate cause, the ultimate support, and the ultimate refuge." (Mundaka Upanishad 1.1.7)

This quote succinctly encapsulates the cosmic significance of Brahman, the ultimate reality. It asserts that all worlds, encompassing the entirety of existence, originate from Brahman, are sustained by Brahman, and ultimately merge back into Brahman. Brahman is depicted as the primordial cause of creation, the foundational support that upholds all of existence, and the ultimate refuge to which all phenomena return. This emphasizes the integral role of Brahman in the cosmic order, highlighting its omnipresence, omnipotence, and eternal nature. Brahman serves as the unifying principle that governs the cyclical process of creation, sustenance, and dissolution, underscoring its paramount importance in the spiritual and philosophical understanding of the universe.

Independence and Self-Luminosity:

- "Brahman is self-existent. Being subtle, He is beyond understanding. He is beyond all differences of form. He is omniscient and is the inner controller." (Mundaka Upanishad 1.1.9)

  Brahman is depicted here as self-existent and beyond comprehension. Described as subtle, Brahman transcends the grasp of ordinary understanding and defies categorization by form. It emphasizes Brahman's omniscience and role as the inner controller, suggesting that Brahman is not only the ultimate reality but also the guiding force within all beings. This portrayal underscores the ineffable nature of Brahman and highlights its omnipotence and transcendence.

- "The supreme Self is beyond name and form, beyond the senses, inexhaustible. Those who realize the Self are forever free from the jaws of death." (Katha Upanishad 1.3.15)

  In this quote, the supreme Self, synonymous with Brahman, is described as transcending all limitations of name and form. Beyond the realm of the senses, it is depicted as inexhaustible and eternal. The realization of the Self is portrayed as the key to liberation from the cycle of birth and death, offering eternal freedom from mortality. This quote underscores the importance of transcending the identification with the finite and recognizing the true essence of the Self, which leads to liberation and eternal bliss beyond the transient realm of existence.

Unchanging Reality:

- "The wise, having realized Brahman as the immutable, eternal reality, attain liberation. Being beyond birth and death, they become immortal." (Chandogya Upanishad 8.12.1)

  Here, realization of Brahman as the unchanging, eternal reality is depicted as the path to liberation. The wise, having attained this realization, transcend the cycle of birth and death, achieving immortality. This quote highlights the transformative power of spiritual wisdom and self-realization, which liberates individuals from the constraints of mortal existence and leads them to eternal freedom.

- "That (Brahman) is infinite, this (universe) is infinite; from the infinite, the infinite arises. When the infinite is taken from the infinite, only the infinite remains." (Isavasya Upanishad 1)

  In this quote from the Isavasya Upanishad, Brahman is described as infinite, encompassing both the universe and beyond. It highlights the concept of the infinite giving rise to the infinite, illustrating the interconnectedness and inseparability of Brahman and the universe. Even when a part of the infinite is seemingly taken away, the infinite remains undiminished. This emphasizes the omnipresence and indivisibility of Brahman, serving as a profound reminder of the ultimate unity and wholeness underlying all of existence.

Source of All Changes and Transformations:

- "That from which all beings originate, by which all beings are sustained, and unto which all beings return — know that to be Brahman." (Taittiriya Upanishad 3.1.1)

  This quote succinctly captures the essence of Brahman as the ultimate reality underlying the cosmos. It emphasizes Brahman as the source of all existence, the sustainer of all beings, and the ultimate destination to which all return. By highlighting the cyclical nature of creation, sustenance, and dissolution, this quote emphasizes the omnipotence and omnipresence of Brahman. It serves as a profound reminder of the interconnectedness of all life and the fundamental unity of existence within the eternal essence of Brahman.

These beautiful passages from the Upanishads illuminate the radiant essence of Brahman, inviting us into the embrace of its timeless wisdom and boundless love. Within these sacred texts, Brahman is portrayed as self-existent, self-sufficient, and infinitely free — a divine presence that transcends all limitations, changes, and dependencies. As we immerse ourselves in the teachings, we are reminded that Brahman stands as the unwavering foundation of existence, the silent witness to the dance of creation and the source of all life.

The descriptions of Brahman as "Unseen, Unbounded, Unconstrained" evoke a sense of awe and wonder, beckoning us to journey beyond the confines of our finite minds and into the realm of spiritual insight. Brahman eludes our grasp, yet its presence is felt in the depths of our being, calling us to awaken to the truth of our interconnectedness with all that is. It is through the silent whispers of intuition, the gentle touch of direct experience, and the profound realization of self that we come to know Brahman — the ineffable essence that animates the universe and resides within each and every one of us.

In contemplating Brahman, we are invited to embark on a sacred pilgrimage — a journey of self-discovery and spiritual awakening that unveils the mysteries of existence and illuminates the path to liberation. With each step we take, we inch closer to the heart of Brahman, where the profound beauty and majesty of reality await, ready to embrace us in its infinite love and wisdom.

How is Brahman different from God or Gods?

How does the idea of Brahman differ from that of God in the Western mindset, or even the conventional Eastern mindset? Are we talking about the same thing but using different names? The short answer to this question, as you can already guess, is no!

While both Brahman and the Western idea of God represent ultimate realities or divine principles, they arise from distinct cultural, philosophical, and theological contexts. In the Western tradition, God is often conceived as a personal deity — a supreme being who possesses attributes such as omniscience, omnipotence, and omnipresence. This God is often depicted as a creator who exists separate from the universe and interacts with humanity through divine revelation, miracles, and providence.

On the other hand, Brahman transcends the limitations of personhood and individual identity. It is not bound by anthropomorphic qualities or relational dynamics but is instead portrayed as an impersonal, all-encompassing reality that underlies and permeates the entire cosmos. Brahman is beyond all forms and attributes, existing beyond the realm of dualities such as good and evil, creator and creation. It is the unchanging, eternal essence that sustains and transcends the flux of phenomenal existence.

While there may be similarities in the underlying concepts of ultimate reality, the cultural and philosophical frameworks within which Brahman and God are understood lead to distinct understandings and interpretations. Therefore, it is not accurate to equate Brahman with the Western concept of God, as they emerge from different theological traditions and philosophical paradigms. Instead, acknowledging the unique qualities and perspectives of each tradition allows for a deeper appreciation of the diversity and richness of the human spiritual experience.

In that spirit, we will draw a more nuanced and critical look at the similarities and differences in the section that follows, albeit on a surface level.

Nature of Ultimate Reality:

- **Brahman:** In Vedanta, Brahman is often understood as the ultimate reality that underlies all of existence. Brahman is described as the unchanging, infinite, and eternal essence from which everything arises and into which everything ultimately merges. Brahman transcends all limitations, including time, space, and causality, and is beyond all attributes, forms, and conceptualizations. Brahman can be perceived as both impersonal and transcendent.
- **God:** In Western theistic traditions, God is typically conceived as a personal deity with distinct attributes, qualities, and characteristics. God is often anthropomorphized, possessing qualities such as intelligence, will, emotions, and the ability to intervene in human affairs. God is

usually perceived as the creator and sustainer of the universe, separate from creation and actively involved in the world.

Non-Duality vs. Dualism:

- **Brahman:** Vedanta, particularly Advaita Vedanta, emphasizes the non-dualistic nature of reality, teaching that Brahman is the only reality and that the individual self (Atman) is identical to Brahman. This non-dualistic perspective suggests that there is ultimately no separation between the individual and the ultimate reality.
- **God:** In Western theistic traditions, there is often a dualistic relationship between God and creation, with God being seen as distinct and separate from the world and individual beings. God is usually portrayed as a creator who stands apart from creation and exercises control over it.

Approach to Worship and Devotion:

- **Brahman:** While Vedanta acknowledges the validity of worship and devotion (Bhakti Yoga) as paths to spiritual awakening, it does not necessarily emphasize a personal relationship with Brahman. Worship and devotion in Vedanta may focus on impersonal aspects of the divine or specific deities within Hinduism.
- **God:** Worship and devotion to a personal God are central components of religious practice in Western theistic traditions. Devotees often cultivate deep personal relationships with God

through prayer, worship, and devotion, seeking divine guidance, protection, and blessings.

Role in Creation:

- **Brahman:** Brahman is often seen as the source and substratum of the entire universe, with everything in creation ultimately arising from and merging back into Brahman. Brahman is not necessarily perceived as an active creator in the same way as the God of theistic traditions.
- **God:** In Western theistic traditions, God is 'usually viewed as the creator and sustainer of the universe, responsible for bringing the world into existence and maintaining order and harmony within it.

So, while there may be some similarities between the concept of Brahman in Vedanta and the concept of God in Western theistic traditions, the two concepts could not be more different in their entirety – especially in their understanding of the ultimate reality, the nature of divinity, the approach to worship and devotion, and the role in creation.

Now that we've explored the concept of Brahman and gained insights into its unique nature, I want to dive into the enchanting idea of Maya. Come with me as we discover the intricate dance between absolute reality and the world of appearances, shedding light on the fascinating interplay that shapes our perception of existence. On this illuminating adventure, we will uncover the mysteries of Maya and its profound connection to the timeless wisdom of Brahman.

A recursive Dialogue:

Who created Brahman?

Brahman is endless and eternal.

What is Brahman made up of?

Brahman is without attribute, it is formless and pure consciousness.

Is Brahman infinite?

Brahman is Absolute and Infinite, not limited by space or time or any object in them.

But, you said Brahman represents Oneness of all creation?

Yes, Brahman is the single binding unity behind diversity in all that exists.

So are infinity and oneness the same?

They are two different ways to realize the Brahman. Brahman is one, not numerically, but in essence. Brahman is the pure unity, be neither diminished, increased, nor divided. And it is eternal and endless. A creation arises from it as a "temporary distortion", just as waves arise from the infinity of the ocean.

Am I Brahman then?

Vedas say, "Aham Bhramasmi," which literally means, I am Brahman. The I here refers to the "atman" or the soul. This signifies the divine, cosmic and timeless nature of our soul and its unity with the "eternal and absolute".

OK, then. If I am Brahman, why do I have all these questions?

Dear fellow traveler of the path, do spend some time meditating not only on the nature of Brahman, but also the nature of maya, and samsara. And then we can talk more!

This is a dialogue I had with one of my senior students in Mexico in 2019 on a rainy afternoon, which I have paraphrased for brevity and clarity.

"Indian philosophers have known, and still know, that Maya, the illusion of separateness and fragmentation, is a condition of our existence in time and space. We must transcend this illusion to see the unity of all things."

– T.S. Eliot

CHAPTER 5

THE DANCE OF ILLUSION: MAYA'S VEIL ON REALITY

Perhaps one of the most quoted concepts of Vedanta in popular folklore is that of Maya!

In fact, growing up in India, and then circulating for most of my life in various spiritual circles, one of the most common phrases I ever heard was, *"It's all Maya."* Sometimes said to highlight the meaninglessness of the rat race of life, at other times it was uttered to imply the futility of attachment and everything in between! This concept also followed me into monkhood, when I lived as a Buddhist monk, where we had an elaborate discourse about Maya.

Interestingly, popular culture also takes a lot of inspiration from Maya. One could argue that "We live in the Matrix," is another euphemism for Maya and the 'blue pill/red pill' conversations in New Age circles.

Maya is, however, actually a very profound philosophical and spiritual concept. To my mind, it remains unmatched in terms of elucidating the impermanent nature of self, lived experience, and the universe itself. Some grounding in that allows us to nurture the inner freedom that has the unique ability to catalyze and accelerate spiritual awareness and awakening.

I could also argue this point through the lens of Existential Philosophy and even Quantum Physics, especially some branches that lean into analogous constructs. Nevertheless, that is arguably outside the scope of this book and I will follow editorial advice and stick to the context of Vedanta!

In Vedanta, then, the concept of Maya refers to the principle of illusion or cosmic ignorance that veils the true nature of reality. Maya is understood as the power or creative energy of Brahman (the ultimate reality) that projects the diverse and manifold universe of names and forms. In some literature, Maya is personified as a goddess who weaves the fabric of the universe and deludes beings into mistaking the transient world of appearances for ultimate reality.

Maya is like a magic show that makes us see things that aren't really there. It's the idea that what we see and experience in the world isn't the whole truth. That there's something deeper beyond what our senses tell us.

When people imagine Maya as a goddess, it is more like she is a character in a story. In this case, Maya represents the force that creates the illusion of the world, as if there's a storyteller weaving a tale that we get caught up in.

When we can see through the illusion and realize the truth about Maya, it's like waking up from a dream and understanding that what you thought was real was just a trick of the mind. By understanding the illusion for what it is, we can find freedom from its grip.

Concerning enlightenment, this is about going beyond what we see and experience in the world and discovering a deeper truth or reality. It's about becoming aware that there's more to life than meets the eye and finding a sense of peace and understanding amidst the complexities of existence.

Let's take some time below to superimpose the concept of Maya on modern-day life so we can gain a clearer understanding of how it manifests itself.

How do we experience Maya?

In the vibrant activity of a bustling city street, Maya weaves its intricate web, captivating our senses and clouding our perception of reality. The towering skyscrapers symbolize the allure of material possessions, enticing us with promises of status and success. As we navigate through the maze of honking cars and bustling pedestrians, the constant distractions of technology pull us further into the illusion, drawing our attention away from the deeper truths of existence.

Maya manifests in myriad forms in this modern landscape, each one a seductive mirage that promises happiness and fulfillment. We may find ourselves ensnared by the pursuit of wealth, believing that financial success holds the key to our well-being. The relentless pursuit of material possessions, driven by societal pressures and cultural expectations, can blind us to the richness of life beyond mere accumulation.

Similarly, the incessant clamor of technology adds another layer to the illusion, pulling us into a virtual world of instant gratification and superficial connections. The allure of social media likes and virtual validation can create a false sense of worth, leading us further away from the deeper reality that lies within.

In the midst of this modern-day Maya, it's easy to become entranced by the surface glitter and glamor, forgetting to look beyond to the deeper truths that lie beneath. Like a mesmerizing illusion, Maya captivates our attention, obscuring the eternal essence that resides within each of us.

Yet, amidst the hustle and bustle of city life, Vedanta offers a timeless message of liberation. Through the practice of self-inquiry, meditation, and spiritual wisdom, we can pierce through the veil of Maya and awaken to the deeper reality of our true nature. Beyond the illusion of success and material wealth, beyond the distractions of technology and superficiality, lies the eternal truth of Brahman – the unchanging essence that transcends all illusions and grants us lasting peace and fulfillment.

How does Maya affect our lived experience?

Like a master illusionist, Maya subtly shapes our thoughts, desires, and actions, leading us down paths that promise fulfillment but ultimately leave us longing for more. It is the fog of ignorance *(Avidya)* that clouds our understanding of ourselves and the world around us, blurring the boundaries between truth and illusion.

In our quest for happiness and satisfaction, Maya beckons us with the seductive allure of fleeting pleasures and external validations. We may find ourselves chasing after material possessions, believing that they hold the key to our happiness. Yet, despite acquiring wealth and status, we are left feeling empty and unfulfilled, as the temporary satisfaction fades away like a mirage in the desert.

Similarly, Maya lures us into the trap of seeking validation from others, measuring our worth by external standards of success and approval. We become ensnared in a cycle of comparison and competition, constantly striving to meet societal expectations and gain recognition from our peers. Yet, no amount of praise or accolades can fill the void within, as Maya perpetuates the illusion that our value lies in the eyes of others rather than within ourselves.

Caught in the web of Maya, we remain blind to the inherent peace and contentment that lie at the core of our being. We are like the thirsty traveler searching for water in the desert, unaware that the oasis of inner fulfillment lies within reach, obscured by the illusions of Maya.

Vedanta offers a beacon of hope amidst the darkness of Maya. Through the practice of self-inquiry, meditation, and spiritual realization, we can see through the veil of illusion. By cultivating inner awareness and discernment, we can transcend the cycle of craving and attachment, finding lasting peace and contentment in the realization of our true nature as divine beings. In the light of Vedanta, Maya's grip loosens, and we are free to walk the path of liberation, guided by the eternal wisdom that shines from within.

Can we liberate ourselves from Maya?

In Vedanta philosophy, liberation (Moksha) stands as the ultimate goal of human existence, beckoning us to transcend the illusions of Maya and awaken to the eternal truth of Brahman. This journey towards liberation is not a mere escape from the clutches of Maya but a profound transformation of consciousness – a journey of self-realization and spiritual awakening.

At the heart of this journey lies the cultivation of spiritual wisdom and self-realization. Through practices such as mindfulness, introspection, and discernment, we embark on a sacred pilgrimage inward, delving into the depths of our being to uncover the timeless truths that lie dormant within. Like a skilled sculptor chiseling away at marble, we chip away the layers of ignorance and illusion that obscure our understanding to gradually reveal the essence of our true nature.

Mindfulness becomes our guiding light, illuminating the path ahead and anchoring us in the present moment. Through the practice of mindfulness, we learn to observe the ebb and flow of thoughts and emotions without attachment or judgment, gaining insight into the transient nature of the phenomenal world. With each moment of mindful awareness, we shred the veils of illusion that shroud our perception, inching closer to the eternal truth that lies beyond.

Introspection becomes our mirror, reflecting the inner landscape of our thoughts, desires, and motivations. Through introspection, we uncover the hidden patterns and conditioning that shape our lives, peeling back the layers of ego and identity to reveal the unchanging essence that lies at the core of our being. With each moment of introspective inquiry, we dissolve the illusions of separation and individuality, recognizing the interconnectedness of all existence.

Discernment helps us to cut through the tangled web of illusion with clarity and insight, helping us learn to distinguish between the transient and the eternal, the ephemeral and the timeless. We see through the seductive allure of Maya, recognizing it for what it truly is — a fleeting mirage that veils the eternal truth of Brahman.

With each act of discernment, we open our hearts and minds to the boundless possibilities of liberation.

Through self-realization, we can transcend the illusions of Maya and awaken to the eternal truth of our divine nature. Freed from the shackles of ignorance, we can glow in the light of liberation, embracing the infinite possibilities of existence with open arms.

Why does Maya even exist?

Think of it this way: if Maya didn't exist, how could we evolve, grow, and reach enlightenment? Maya emerges as the divine play *(Lila)* orchestrated by Brahman — the ultimate reality — for the profound purpose of self-discovery and spiritual evolution. Like a masterful stage director, Maya sets the scene for the grand drama of existence, inviting us to embark on a journey of awakening and self-realization.

At its core, Maya serves as the cosmic canvas upon which the drama of life unfolds. It provides the backdrop against which we navigate the complexities of existence, encountering challenges, triumphs, and moments of profound insight. Maya is not merely an obstacle to be overcome but a sacred catalyst for growth and transformation, prompting us to question the nature of reality and our place within it.

Maya challenges us to look beyond the surface appearances of the material world and peer into the depths of our own being. It beckons us to explore the mysteries of existence, inviting us to delve into the timeless truths that lie hidden beneath the veil of illusion. In the midst of life's trials and tribulations, Maya serves as the impetus for us to seek the wisdom and insight necessary for spiritual evolution.

It's through the lens of Maya that we come to understand the limitations of the material world – that they are but a fleeting illusion, obscuring the eternal truth of our divine nature. Maya prompts us to awaken to our true identity as divine beings, transcending the boundaries of the ego and recognizing our interconnectedness with all of creation.

In this divine play of Maya, every scene becomes an opportunity for self-discovery and spiritual growth. As we navigate the twists and turns of life, we learn to embrace the inherent challenges and uncertainties with grace and equanimity. Maya becomes our greatest teacher, guiding us along the path of self-realization and leading us toward the ultimate goal of liberation (Moksha).

Ultimately, Maya serves as a reminder of the eternal nature of existence, where the boundaries between the self and the other dissolve into the boundless expanse of Brahman. Through Maya, we discover the infinite potential that lies within us. And in this realization, we find liberation — not from Maya itself, but from the illusion of separation, unveiling the eternal truth that we are one with the ultimate reality, Brahman.

If it's all Maya, then what's the point of anything, including enlightenment?

However distracting it may seem, Maya ultimately emerges not as a barrier to enlightenment, but as a sacred catalyst for spiritual growth and self-realization. While it may veil the ultimate reality of Brahman, it simultaneously serves as the fertile ground upon which the seeds of enlightenment are sown.

Enlightenment, in the Vedantic tradition, is not a journey of escapism or renunciation, but rather a journey of transcendence — a journey that involves piercing through the illusions of Maya and awakening to the deeper truth of existence. It is a recognition that the pursuit of spiritual awakening is not about turning away from the world, but about embracing it fully and seeing beyond its transient nature.

Within the realm of Maya, every experience, every interaction, becomes a sacred opportunity for growth, learning, and self-discovery. Whether we find ourselves immersed in moments of joy or grappling with the challenges of life, each encounter offers a doorway to deeper understanding and insight. Maya becomes our greatest teacher, guiding us along the path of self-discovery and leading us toward the realization of our true nature.

Enlightenment is the recognition that amidst the ever-changing flux of Maya, there exists a timeless essence — the unchanging reality of Brahman — that infuses meaning and purpose into every moment of our lives. It is the understanding that beyond the illusions of the material world, there lies a deeper truth — a truth that transcends the limitations of time and space.

Ultimately, enlightenment is not the attainment of some distant goal, but a recognition of the divine essence that resides within each and every one of us. It is the realization that we are not separate from the ultimate reality of Brahman, but are, in fact, an integral part of its eternal unfolding.

Just as we did in the previous chapter(s), let me now share with you exactly what is said about "Maya" directly from the Upanishads and related texts. Perhaps we can pierce a small hole through the Maya about Maya! ☺

Veil on Reality:

Maya is described as a veil that obscures the true nature of reality, concealing the underlying unity and wholeness of existence. It creates the illusion of diversity, multiplicity, and separation in the world, leading beings to perceive themselves as distinct and separate from one another and the ultimate reality (Brahman).

According to the Upanishads, the nature of Maya is portrayed as:

A Veil Obscuring Reality:

- "Maya is the inscrutable power of Brahman. By Maya, Brahman appears to be divided into infinite objects and beings." (Svetasvatara Upanishad 4.10)

  This quote encapsulates the profound concept of Maya in Vedanta philosophy. It suggests that Maya is the mysterious and incomprehensible power of Brahman, the ultimate reality. Through Maya, Brahman appears to be fragmented into an infinite array of objects and beings, creating the illusion of diversity and multiplicity in the phenomenal world. In other words, Maya is the divine force that veils the underlying unity of Brahman and gives rise to the illusion of separation and diversity in the universe.

The Illusion of Diversity and Multiplicity:

- "As one fire, after it has entered the world, though one, takes different forms according to whatever it burns, so does the internal Atman of all living beings, though one, takes a form according to whatever It enters, and exists also without." (Brihadaranyaka Upanishad 2.4.6)

  This quote uses the metaphor of fire to illustrate the nature of the Atman, the inner self or soul, in relation to all living beings. Just as a single fire, upon entering the world, may take on different forms depending on what it burns, the Atman, though inherently one, assumes various forms according to the bodies it inhabits. However, the essence of the Atman remains unchanged and exists independently of these forms. In essence, it emphasizes the unity of the Atman across all beings, despite the diversity of individual manifestations.

The Perception of Separation from Ultimate Reality:

- "As water poured into water, milk into milk, ghee into ghee become one and the same, so does the Atman of the wise become one with Brahman." (Chandogya Upanishad 6.8.7)

  Here, the Chandogya Upanishad employs the analogy of various substances merging seamlessly to illustrate the concept of unity between the Atman (individual self) and Brahman (ultimate reality). Just as water poured into water, milk into milk, and ghee into ghee blend completely, losing their individual identities and becoming indistinguishable from each other, similarly, the Atman of the wise merges with Brahman. This analogy underscores the idea that the enlightened individual recognizes their essential identity with the ultimate reality, transcending the illusion of separateness and realizing their inherent oneness with Brahman.

Overcoming the Illusion of Maya:

- "Maya, the divine enchantress, the veil that covers knowledge, throws the individual into the ocean of samsara (worldly existence), blinding them to their true nature." (Katha Upanishad 1.3.11)

  Maya is depicted as a captivating force that blinds individuals to their inherent divinity and leads them astray from realizing their true nature. Like a mesmerizing spell, Maya deludes individuals into identifying with the ephemeral and transient aspects of existence, preventing them from recognizing their eternal essence. Through this imagery, the Upanishad highlights the pervasive and deceptive nature of Maya, emphasizing the importance of transcending its illusion to attain liberation and self-realization.

- "When one has realized Brahman, which is hidden in Maya, then by Maya one does not become deceived." (Katha Upanishad 2.3.14)

  This quote suggests that upon attaining the realization of Brahman, the ultimate reality hidden within Maya, one transcends the deceptive nature of Maya itself. In other words, through the direct experience of Brahman, individuals gain insight into the illusory nature of Maya and are no longer subject to its deceptions. This implies that spiritual enlightenment grants individuals the discernment and clarity needed to see through the veils of illusion and perceive the underlying truth of existence. By recognizing the inherent unity of Brahman within Maya, individuals become immune to its illusions and attain a state of unshakeable awareness and liberation.

In the enchanting world of Vedanta, Maya is the elusive veil that conceals the essence of reality. The profound insights to be found in the above quotes paint Maya as more than just a veil — it's a captivating enchantress, weaving illusions of diversity, multiplicity, and separation in the fabric of existence. Yet, amidst this cosmic illusion, there lies a path of spiritual awakening and understanding. By embracing spiritual awareness, seekers can pierce through the veil of Maya, revealing the hidden unity and interconnectedness that underpins all of existence. It's an invitation to embark on a journey of self-discovery and enlightenment, where the mysteries of Maya unravel to unveil the boundless wholeness and oneness of the universe.

Illusion of Duality:

Maya is responsible for the appearance of duality and opposites in the world, such as good and evil, pleasure and pain, birth and death, and so on. It creates the illusion of distinctions, divisions, and limitations, causing beings to experience the world through the lens of dualistic perception.

How do the Upanishads present this illusion? Let's take a look at some quotes from the ancient texts:

Maya as the Source of Duality:

- "Maya covers the knowledge of reality, causing the world to appear as if separate from Brahman, just as a cloud covers the sun." (Svetasvatara Upanishad 4.10)

  This quote employs the metaphor of a cloud covering the sun to convey the idea that Maya veils our understanding of Brahman, much like how a cloud obscures the brilliance of the sun. Maya creates the illusion of separation between the individual and the ultimate reality and leads us to perceive the world as fragmented and disconnected from the divine source. However, just as the sun remains ever-present behind the cloud, Brahman is always present beneath the veil of Maya, waiting to be revealed through spiritual insight and realization.

Through this analogy, the quote emphasizes the transient and illusory nature of Maya and encourages seekers to cultivate spiritual awareness and discernment to see beyond the illusions and recognize the underlying unity that permeates all of existence.

Illusion of Opposites:

- "This world is an illusion, the other side of the truth. It has both the real and the unreal, and leads to the truth." (Mundaka Upanishad 2.2.6)

  This quote suggests that the world is a mixture of reality and illusion, serving as a pathway to understanding the ultimate truth. It implies that although the world may appear transient and unreal, it also contains elements of truth that can lead seekers toward a deeper understanding of reality. In essence, the world is a realm where both the real and the unreal coexist, ultimately guiding individuals toward enlightenment and truth.

Distinctions and Limitations:

- "Maya creates the illusion of diversity and separation, causing beings to perceive themselves as limited and separate entities in the world." (Chandogya Upanishad 6.9.2)

  Maya is depicted as a force that generates a deceptive perception of diversity and separation. It suggests that Maya veils the underlying unity of existence, leading beings to mistakenly view themselves as isolated and finite entities within the world. This illusion of separateness prevents individuals from recognizing their inherent connection to the broader reality and obscures the fundamental oneness that underlies all existence.

Perception through Dualistic Lens:

- "Bound by the fetters of Maya, people see not the truth." (Mundaka Upanishad 2.2.5)

  Maya is portrayed as a binding force that restricts people from perceiving the truth. It suggests that individuals, ensnared by the illusions created by Maya, are unable to discern the ultimate reality. The metaphor of being bound by fetters implies a state of bondage or limitation, indicating that Maya obstructs the clear perception of truth and veils the deeper understanding of existence.

We can see clearly how Maya is responsible for the appearance of duality and opposites in the world, such as good and evil, pleasure and pain, birth and death, and so on. It creates the illusion of distinctions, divisions, and limitations, causing beings to experience the world through the lens of dualistic perception.

Role in Creation:

Maya is also understood as the creative power of Brahman that brings forth the universe into existence. It is through Maya that Brahman manifests as the diverse and multifaceted world of names and forms, including all living beings, objects, and phenomena.

Maya as the Creative Power of Brahman:

- Maya is like the magic wand of Brahman, bringing dreams into reality. It's the creative force that makes everything in the universe come alive.

  In this metaphorical interpretation, Maya is likened to a magical wand wielded by Brahman, the ultimate reality. It symbolizes the creative power that transforms dreams into tangible reality, breathing life into the universe. This portrayal suggests that Maya is the dynamic force behind the manifestation of all phenomena, animating the cosmos and bringing forth the diverse forms and experiences of existence. It highlights Maya's role as the creative energy that infuses vitality and vibrancy into the fabric of reality, shaping the ever-changing tapestry of the universe.

- "This Maya is hard to pierce, O Brahman, save by those who have reached the other shore." (Mundaka Upanishad 1.2.8)

Manifestation of the Diverse Universe:

- Maya weaves the tapestry of existence, turning Brahman's potential into reality. It's like the architect who designs and constructs the magnificent palace of the universe.

  In this depiction, Maya is envisioned as the master weaver who intricately crafts the fabric of existence, transforming the latent potential of Brahman into tangible reality. Similar to an architect designing and constructing a grand palace, Maya meticulously shapes and constructs the vast and magnificent universe. This imagery illustrates Maya's role as the creative force that orchestrates the myriad forms and phenomena of the cosmos, imbuing each aspect of existence with beauty, complexity, and harmony. Through this lens, Maya emerges as the divine artisan who brings Brahman's infinite possibilities to fruition, weaving together the threads of creation to compose the rich and diverse tapestry of reality.

- "From Brahman all beings originate, by Brahman they are sustained, and unto Brahman they return. Brahman is the truth, Brahman is the Self, and that you are." (Mundaka Upanishad 2.2.10)

  This passage invites us to recognize the dynamic interplay between Brahman, Maya, and the manifested universe. It encourages us to embrace the diversity and complexity of existence while also acknowledging the underlying unity that connects all things back to the divine source. Through this perspective, Maya is not seen as separate from Brahman but rather as an integral part of the divine unfolding of creation.

Creation of Living Beings and Phenomena:

- Maya is the cosmic artist who sculpts every blade of grass, every star in the sky, and every beating heart. It's the brushstroke that paints the colors of life onto the canvas of existence.

Maya is the divine creative power that transforms the infinite potential of Brahman into the manifest world of names and forms. It's the mechanism through which Brahman expresses itself as the diverse and multifaceted universe, encompassing all living beings, objects, and phenomena.

Obstacle to Spiritual Realization:

Maya is considered a fundamental obstacle to spiritual awakening and liberation (moksha). It deludes beings into identifying with the transient world of appearances, leading them to become entangled in the cycle of birth and death (samsara). Overcoming Maya and realizing the true nature of reality is the central goal of spiritual practice in Vedanta.

Maya as a Fundamental Obstacle:

- Maya acts like a mirage in the desert, leading travelers astray. It's the fog that clouds the mind and veils the truth, preventing individuals from recognizing their true nature. Like a metaphorical fog, Maya obscures clarity of thought and hides the truth, thereby hindering individuals from realizing their authentic selves.
- "Maya, the divine enchantress, the veil that covers knowledge, throws the individual into the ocean of samsara (worldly existence), blinding them to their true nature." (Katha Upanishad 1.3.11)

Here again, Maya is personified as a divine enchantress, symbolizing its captivating and beguiling nature. It is a veil that shrouds true knowledge, leading individuals into the vast ocean of samsara, or worldly existence. By doing so, Maya blinds individuals to their innate essence and divinity, preventing them from realizing their true nature and perpetuating their entanglement in the cycle of birth, death, and rebirth.

Entanglement in the Cycle of Birth and Death:

- Maya is like the spinning wheel of fate, trapping souls in the endless cycle of birth, death, and rebirth. It's the web of desires and attachments that bind individuals to the transient world of appearances.

  In other words, Maya is the cosmic force that perpetuates the illusion of individuality and separation, leading souls away from their true nature and trapping them in the cycle of samsara. By transcending the web of Maya's illusions through spiritual insight and realization, we can break free from the cycle of birth and death and attain liberation from the bondage of worldly existence.

Central Goal of Spiritual Practice:

- Overcoming Maya is like breaking free from chains and stepping into the light of truth. It's the journey of self-discovery and awareness, leading individuals to recognize their essential unity with Brahman.

- "When one has realized Brahman, which is hidden in Maya, then by Maya one does not become deceived." (Katha Upanishad 2.3.14)

  This quote also suggests that upon attaining realization of Brahman, the ultimate reality hidden within Maya, individuals are no longer deceived by Maya's illusions. It implies that spiritual enlightenment grants individuals the discernment and clarity needed to see through the veils of illusion and perceive the underlying truth of existence.

In Vedanta, the ultimate goal of spiritual practice is to transcend the illusion of Maya and perceive the true nature of reality. It's the journey of awakening from the dream of separation and recognizing the underlying unity and wholeness of existence. Through self-inquiry, meditation, and devotion, individuals seek to overcome Maya and attain liberation (moksha) from the cycle of birth and death.

Transient Nature:

Maya is characterized by its transient and ephemeral nature. It is not an absolute reality but a temporary phenomenon that arises and subsides within the ultimate reality of Brahman. While Maya creates the appearance of diversity and change, Brahman remains unchanged and unaffected by the play of Maya. Maya is like a passing cloud in the sky, constantly shifting and changing. It's the fleeting mirage in the desert, here one moment and gone the next. Maya is impermanent and ever-changing, like the waves on the surface of the ocean.

Transient and Ephemeral Nature:

- "This world is an illusion, the other side of the truth. It has both the real and the unreal, and leads to the truth." (Mundaka Upanishad 2.2.6)

Not an Absolute Reality:

- Maya is like the reflection in a mirror, appearing real but lacking substance. It's the dream that fades upon waking, revealing the true reality beyond. Maya is not an absolute reality but a temporary phenomenon that arises and subsides within the ultimate reality of Brahman.
- "When the seer sees the maker, who is difficult to see, hidden in Maya, then the seer rejoices, for he has perceived the supreme reality." (Svetasvatara Upanishad 4.14)

  In this quote, Maya is depicted as the veil that obscures the sight of the supreme reality, Brahman. However, when a seer — a spiritually enlightened individual — successfully perceives the divine presence hidden within Maya, there is a profound sense of joy and fulfillment. This experience of recognizing the ultimate reality amidst the veils of illusion is considered a moment of supreme realization and enlightenment. It signifies the seer's deep understanding and connection with the underlying truth of existence, which transcends the limitations imposed by Maya.

Appearance of Diversity and Change:

- Maya is like the kaleidoscope of existence, constantly shifting colors and patterns. It creates the illusion of diversity and change, leading beings to perceive the world as a multiplicity of forms and phenomena.
- "From Brahman all beings originate, by Brahman they are sustained, and unto Brahman they return. Brahman is the truth, Brahman is the Self, and that you are." (Mundaka Upanishad 2.2.10)

  This quote encapsulates the central teachings of Vedanta philosophy, emphasizing the inseparable connection between the individual self and the supreme reality of Brahman. It invites individuals to recognize their essential identity with Brahman and to realize the ultimate truth of their existence.

Unchanged and Unaffected Brahman:

- Amidst the ever-changing play of Maya, Brahman remains unchanged and unaffected. It's like the stillness at the center of a spinning wheel, the unchanging reality amidst the flux of appearances. This understanding invites individuals to transcend the ephemeral realm of Maya and seek refuge in the timeless truth of Brahman.

The Upanishads convey a profound message about Maya, depicting it as the veil of illusion that obscures the true nature of reality. While Maya engenders diversity, multiplicity, and separation in the world, it ultimately veils the underlying unity and oneness of existence. Through various metaphors and analogies, the Upanishads emphasize Maya's transient and illusory nature, cautioning individuals against becoming ensnared in its deceptive web. However, amidst Maya's play, Brahman remains unchanged and unaffected, serving as the enduring truth beyond the realm of appearances. The overall message of the Upanishads about Maya encourages seekers to transcend its illusions through spiritual insight and realization, thereby unveiling the deeper reality of Brahman and attaining liberation from the cycle of worldly existence.

The concept of Maya and its veil of illusion serves as a profound metaphor for understanding the nature of reality and the human condition. It invites us to look beyond the surface of appearances and to penetrate the mirage in order to grasp the underlying unity and truth of existence. Through self-inquiry, meditation, and spiritual practice, we can overcome Maya and attain spiritual liberation, realizing our essential unity with Brahman.

In the final chapter of this section, we will delve into the concept of the individual self or Atman. As a taster, I want to mention that Atman represents the true essence of the individual, transcending the illusions of Maya and reflecting the eternal reality of Brahman within. Let's explore the profound insights offered by Vedanta on the nature of the self, taking you on a transformative path that can lead to spiritual liberation and ultimate realization of the timeless truth that resides within.

Munch and Masala

If I am one with you and you are one with me why are we so different?

Because it allows us to experience each other and learn from each other.

If we are one, then we already know. What else is there to learn?

Learning is also purification, transmutation of karma through awakening.

If all is one, then I am already awakened and so are you. Why this karma?

It's the dance of maya, eternal and yet impermanent.

So am I Maya?

All is Maya.

Who is striving to awaken, then?

The one who needs to remember he is awakened.

Now it makes sense. Thank you.

Aham Brahmasmi!

A conversation I had with a student during a retreat in Vienna in 2016.

"The Atman is the soul of man, the universal self, of which every individual soul is a manifestation. It is the essence of all beings, the substratum of everything."

– Henry David Thoreau

CHAPTER 6

SACRED SELF: EXPLORING THE DEPTHS OF ATMAN

After over three decades of studying different narratives within the universe of spirituality, ranging from Eastern mysticism to practices of the ancient Egyptians, one theme or archetype has always stood out to me as a common thread. That is the concept of a soul or *alma, manna, dusa, ziel,* and many other words referring to the soul across cultures, civilizations, and languages.

While there is a great deal of "usage" of this word, the essential meaning or archetype of *a life force* varies widely. From transcendental qualities being ascribed in Eastern systems to an "image of god" in some Western faiths, the term "life force" has many interpretations. One of the first points I want to make clear is that when a Catholic, Hindu, Muslim, Jew, Sufi, Peruvian Shaman, or philosopher of Metaphysics are talking about the Soul, they are not referring to the same "essential meaning". Especially in texts written in English, a diverse range of archetypes get bunched under the catchall of the word *Soul*. So let's demystify the word a little bit by digging deeper into what the Soul means in the Vedic context!

In Vedanta, we have the captivating concept of the Sacred Self, known as Atman. Atman isn't just your ordinary sense of self — it's the very essence that connects us to the divine, the ultimate reality known as Brahman. Picture it as the eternal flame burning within every living being, transcending the bounds of the physical body and mind. It's our core identity, our consciousness, and often gets lost in translation as just a "soul" in English.

But here's where it gets truly fascinating — according to Vedanta, Atman isn't separate from Brahman; it's one and the same. Think of it like a drop of water seamlessly merging with the vast ocean, losing its individual identity in the process. Realizing Atman's true nature is the holy grail of spiritual seekers — it's the key to unlocking liberation from the relentless cycle of birth and death, known as moksha.

In Christianity, Judaism, and Islam, the soul is often viewed as an immortal and incorporeal essence that represents the spiritual aspect of human beings. It is believed to be created by a divine entity (God) and is distinct from the physical body. The soul is usually associated with concepts such as morality, conscience, and personal identity, and it is believed to continue to exist after physical death. Depending on the specific religious tradition, the fate of the soul after death may involve concepts such as heaven, hell, or purgatory.

In contrast, Atman is perceived as eternal and infinite, free from the constraints of birth and death that characterize such ideas of the soul. While the soul in Western traditions may face judgment and consequences in an afterlife based on moral actions, Atman transcends such earthly concerns, basking in the eternal glow of infinite existence.

Furthermore, the concept of Atman places a strong emphasis on self-realization and the uncovering of one's true nature through spiritual practices like meditation and self-inquiry. In contrast, Western notions of the soul may prioritize moral behavior and adherence to religious doctrines as a means of ensuring salvation or avoiding damnation.

On the other hand, Eastern Buddhism teaches the concept of "Anatta," which translates to "non-self" or "no-soul." Instead of a permanent and unchanging soul,

Buddhism emphasizes the idea of "Anicca" (impermanence) and "Dependent Origination," suggesting that what we perceive as the self is a transient combination of elements. Enlightenment is attained by realizing the true nature of existence.

While Vedanta teaches that Atman is eternal and unchanging, existing beyond the cycle of birth and death, Buddhist teachings emphasize that all phenomena, including the self, are subject to constant change and devoid of any permanent essence. It stresses that enlightenment is attained by realizing the true nature of existence, which involves understanding the impermanence, suffering, and non-self nature of all phenomena. Equally, while Vedanta places a strong emphasis on the authority of the Vedas and Upanishads, about the nature of Atman, Buddhism rejects the authority of the Vedas and emphasizes the teachings of the Buddha, as found in the Pali Canon and other Buddhist scriptures.

In Taoism, the concept of the soul is not explicitly defined but is often associated with the idea of "Qi" (life force energy) and the interconnectedness of all things. Taoist practices aim to cultivate harmony with the natural flow of Qi and achieve spiritual balance. This differs from the Vedantic concept of Atman, which is believed to be eternal, unchanging, and the essence of individual identity.

While Taoist teachings emphasize the interconnectedness of all things, with Qi flowing through and connecting every aspect of existence, Vedanta teaches that Atman is not only interconnected with all beings but is identical with Brahman, the ultimate reality. This concept of non-duality (Advaita) suggests that individual souls are not separate from the divine but are manifestations of the same ultimate reality. In addition, Taoist practices to cultivate harmony with the natural flow of Qi and achieve spiritual balance include qigong, tai chi, and meditation, as opposed to the Vedantic practices of self-inquiry, meditation, and devotion.

Many indigenous cultures have beliefs in the existence of a soul or spirit that animates living beings and connects them to the natural world. These beliefs vary widely depending on the specific cultural traditions and may involve rituals, ceremonies, and practices aimed at honoring and communing with ancestral spirits or nature spirits. Again, this contrasts with the concept of Atman in Vedanta, which views the individualized self as identical to Brahman, the ultimate reality.

Although indigenous beliefs often emphasize the interconnectedness between humans, nature, and the spiritual realm, Vedanta emphasizes the identity of Atman with Brahman, rather than specific spirits or entities within nature. Also, while indigenous cultures may have elaborate rituals and ceremonies dedicated to honoring and communing with spirits, ancestors, and the natural world, Vedantic practices focus on spiritual disciplines aimed at realizing the true nature of Atman and achieving liberation from the cycle of reincarnation.

In various traditional African religions, there are beliefs in ancestral spirits that continue to exist after death and can influence the lives of the living. These spirits are often revered and consulted through rituals and ceremonies. In some ways, this is similar to the existence of Atman, the individualized self, which is considered eternal and unchanging, existing beyond the cycle of birth and death. But that's where the similarities end. Traditional African religions view ancestral spirits as distinct entities that maintain a connection with their descendants and the living world. In contrast, the concept of Atman in Vedanta emphasizes that Atman transcends individual identity and merges with the divine.

While both Atman and the soul explore the essence of individual identity across cultures and belief systems, Atman in Vedanta offers a more expansive and inclusive understanding. It emphasizes unity with the divine and liberation from the cycle of worldly existence.

So how do we uncover this divine essence within us? It's all about self-inquiry, meditation, and delving deep into the realms of self-exploration and introspection. Through these profound practices, we embark on an exhilarating journey to unravel the mysteries of Atman, unveiling its eternal and boundless nature as the very essence of our existence.

Here's a closer look at the exploration of the depths of Atman, directly from Vedic texts:

Essence of Being:

Atman is understood as the essence of being, the true self that transcends the limitations of the body, mind, and ego. It is described as eternal, unchanging, and indestructible, existing beyond birth and death. Atman is considered the divine spark that animates life and consciousness.

Let's dig deeper into the nature of Atman, as described in the Upanishads:

Atman as the Essence of Being:

- "The Self (Atman) is the inner controller, the source of all desires, the eternal witness, the ever-present reality amidst the changing phenomena of the world." (Brihadaranyaka Upanishad 3.7.15)

  This quote highlights the essence of the Self, or Atman, according to Vedanta philosophy. It describes Atman as the inner controller, indicating its role as the source of individual agency and the seat of consciousness within each being. Atman is portrayed as the source of all desires, signifying that it is the driving force behind human motivations and actions. Furthermore, it is described as the eternal witness, suggesting that Atman is the unchanging observer of all experiences, untouched by the transient fluctuations of the world. Finally, Atman is depicted as the ever-present reality amidst the changing phenomena of the world, emphasizing its eternal and unchanging nature in contrast to the impermanence of the external world.

- "Atman is the eternal, immutable, and self-luminous essence within every living being. It is the witness of all experiences and the source of consciousness." (Brihadaranyaka Upanishad 2.4.12)

  Here, we find a profound insight into the nature of Atman, the innermost self according to Vedanta philosophy. Atman is described as eternal, indicating its timeless and unchanging nature that transcends the limitations of the material world. It is further characterized as immutable, suggesting that Atman remains constant and unaffected by the fluctuations of existence. Additionally, Atman is portrayed as self-luminous, implying that it radiates its own intrinsic light of consciousness, illuminating the inner world of every living being. Furthermore, Atman is depicted as the witness of all experiences, signifying its role as the silent observer that perceives and comprehends the unfolding of life's events.

Eternal and Unchanging Nature:

- "The Self (Atman) is not born, nor does It die. It did not originate from anything, nor did anything originate from It. It is unborn, eternal, and everlasting. It is not slain when the body is slain." (Katha Upanishad 1.2.18)

  In the Katha Upanishad, we learn of the timeless understanding of Atman, the innermost self in Vedanta philosophy. Atman is portrayed as beyond the realm of birth and death, transcending the cycle of existence. It is described as unoriginated and eternal, suggesting that Atman has always existed and will continue to exist indefinitely. Unlike the transient nature of the physical body, Atman remains unaffected by the process of birth and death, existing beyond the limitations of time and space. Even when the physical body perishes, Atman persists, undiminished and everlasting.

Indestructible Essence:

- "That which is the finest essence - this whole world has that as its soul. That is Reality. That is Atman. That art thou." (Chandogya Upanishad 6.8.7)

  This verse encapsulates the essence of Atman, portraying it as the fundamental essence of the universe and the ultimate reality. It emphasizes Atman's omnipresence, identifying it as the soul of the world and the essence of all beings. The verse concludes with the powerful declaration, "That art thou," highlighting the unity between the individual self and the universal self, inviting the seeker to realize their inherent divinity within the fabric of existence.

Innermost Essence and Divine Spark:

- "The wise man perceives unity in the midst of diversity, whereas the ignorant sees diversity only." (Isavasya Upanishad 6)

  In this quote, we see a comparison between the perspectives of the wise and the ignorant regarding the nature of reality. It suggests that the wise person perceives unity within the diverse manifestations of the world, recognizing the underlying interconnectedness and harmony. In contrast, the ignorant individual sees only diversity, unable to grasp the deeper unity that pervades all existence. This verse underscores the importance of spiritual insight in transcending apparent differences and realizing the underlying oneness of all creation.

In effect, Atman is understood as the true self that transcends the limitations of the body, mind, and ego. It's described as eternal, unchanging, and indestructible, existing beyond birth and death.

Unity with Brahman:

Vedanta teaches that Atman is identical with Brahman, the ultimate reality. While Atman is the individualized expression of Brahman, it is ultimately of the same essence as Brahman. Realizing this identity between Atman and Brahman is the key to spiritual liberation (moksha) in Vedanta philosophy.

The sacred texts present the unity of Atman with Brahman in several ways:

Identity of Atman with Brahman:

- "The knower of Brahman becomes Brahman. He becomes all this, whatever exists." (Chandogya Upanishad 7.24.1)

  This quote highlights the transformative power of knowing Brahman, the ultimate reality. It suggests that the one who attains true knowledge of Brahman transcends individual identity and merges with the divine essence. In doing so, one becomes imbued with Brahman's infinite nature, encompassing all existence.

- "That art thou." (Chandogya Upanishad 6.8.7)

  This succinct statement encapsulates the profound truth of Vedanta philosophy. It emphasizes the identity between the individual self (Atman) and the universal self (Brahman). "That" refers to the ultimate reality, Brahman, while "thou" signifies the individual self, Atman. This declaration underscores the fundamental unity and oneness of all existence, inviting the seeker to realize their inherent divinity and recognize the divine essence within themselves and in all aspects of creation.

Individualized Expression of Brahman:

- "From Brahman all beings originate, by Brahman they are sustained, and unto Brahman they return." (Mundaka Upanishad 2.2.10)

  Here, we find a summary of the Vedantic perspective on existence. The extract states that all beings originate from Brahman, are sustained by Brahman, and ultimately return to Brahman. Brahman is, indeed, the source, sustainer, and ultimate destination of all existence.

Key to Spiritual Liberation:

- "By knowing the Self (Atman), one transcends all suffering and attains supreme peace." (Mundaka Upanishad 3.1.6)

  Another quote from the same source, it highlights the transformative power of self-realization and suggests that by knowing the true nature of the Self (Atman), one can transcend all suffering and attain supreme peace. This emphasizes the importance of inner wisdom and self-awareness in achieving liberation from the cycles of suffering and finding lasting peace and fulfillment.

Union with Brahman:

- Understanding the identity between Atman and Brahman leads to spiritual liberation (moksha), where the individual soul merges with the ultimate reality, experiencing the state of oneness.

  Embarking on the journey of self-discovery in Vedanta, one can uncover the profound connection between the individual soul (Atman) and the cosmic reality (Brahman). This realization dissolves the illusion of separation, leading to a profound sense of oneness with the universe.

Through practices like meditation and contemplation, seekers merge with Brahman, experiencing liberation and boundless bliss beyond words. Becoming aware of the connection between Atman and Brahman leads to a profound sense of unity and oneness with all existence.

Inward Journey:

Exploring the depths of Atman involves an inward journey of self-discovery and self-realization. It requires looking beyond the surface level of the ego and the mind to uncover the deeper layers of consciousness and awareness.

Here's how the Upanishads present that inward journey:

Inward Journey of Self-Discovery:

- "Know the Self (Atman) to be the master of the chariot, the body to be the chariot, the intellect (buddhi) the charioteer, and the mind (manas) the reins." (Katha Upanishad 1.3.3)

  The Katha Upanishad uses the metaphor of a chariot to explain the relationship between the individual and their faculties. Here, the "Self" (Atman) is likened to the master of the chariot, representing the true essence or consciousness of the individual. The "body" is compared to the chariot itself, serving as the vessel through which experiences occur. The "intellect" (buddhi) is depicted as the charioteer, responsible for guiding and directing the chariot. Lastly, the "mind" (manas) is described as the reins, symbolizing the control and regulation of thoughts and emotions. This analogy illustrates the interconnectedness and interplay between different aspects of the self in navigating the journey of life.

Looking Beyond the Ego and Mind:

- "By the mind alone is the Self (Atman) to be realized. There is no other way." (Brihadaranyaka Upanishad 4.4.6)

  This quote underscores the importance of the mind in realizing the true nature of the Self (Atman). It emphasizes that self-realization can only be achieved through the mind, suggesting that introspection, contemplation, and inner awareness are essential for understanding one's true identity. The mind is the primary tool for transcending ignorance and attaining spiritual enlightenment, highlighting the significance of mental discipline and introspective practices on the path to self-realization.

Means of Exploration:

- "Knowing the self (Atman) to be beyond all sense-perception, inconceivable, immutable, and imperishable, one attains eternal peace." (Katha Upanishad 1.2.24)

  This text suggests that true understanding of the Self goes beyond ordinary perception and conceptualization. The Self is described as beyond the grasp of the senses, implying that it cannot be fully comprehended through sensory experience alone. Instead, it is depicted as inconceivable, immutable, and imperishable, indicating its eternal and unchanging nature. By recognizing the Self in this way, one attains a state of eternal peace, with true liberation coming from transcending the limitations of the material world and realizing the eternal essence within.

Realization of True Nature:

- "The knower of Atman transcends sorrow, transcends death. By knowing Atman, one attains immortality." (Mundaka Upanishad 3.2.9)

  This quote emphasizes the transformative power of knowing the Self (Atman) and suggests that true understanding of Atman leads to liberation from sorrow and the cycle of death and rebirth. By transcending the limited identification with the body and mind, the knower of Atman achieves immortality, indicating a state of eternal existence beyond the fluctuations of the material world.

There is a profound significance of self-realization in attaining liberation and experiencing the eternal nature of the Self. By using practices such as meditation, self-inquiry, and contemplation, one can delve into the depths of consciousness and awareness that ultimately lead to the understanding of Atman's true nature.

Beyond Dualities:

Atman is described as beyond all dualities and opposites, including those of pleasure and pain, good and evil, birth and death. It is the unchanging and eternal essence that remains unaffected by the fluctuations of the phenomenal world. It's like the still center of a spinning wheel, unaffected by the whirlwind of pleasure and pain, good and evil, birth and death.

Transcending Dualities:

- "That which is the finest essence - this whole world has that as its soul. That is Reality. That is Atman. That art thou." (Chandogya Upanishad 6.8.7)

  Again, we read of the unity of the individual self (Atman) with the ultimate reality (Brahman). The essence of the entire world is inseparable from Atman, representing the underlying truth and existence of everything. The phrase "That art thou" reinforces the identity between the individual self and the universal reality, highlighting the interconnectedness of all beings and the inherent divinity within each individual.

Unchanging and Eternal Essence:

- "That supreme Brahman is eternal, and this universe, created by Brahman, is also eternal. Brahman is the cause of the universe, and the universe is the effect of Brahman." (Mundaka Upanishad 2.1.1)

  We learn of the eternal nature of Brahman, the ultimate reality, and the universe it creates. The quote suggests that Brahman is both the cause and effect of the universe, emphasizing the inseparable relationship between the two. This perspective highlights the idea that the universe emanates from Brahman and exists within it eternally.

Realization of True Nature:

- "When all has become the Self (Atman), what should one know and whom should one know by what?" (Brihadaranyaka Upanishad 2.4.14)

  This is a profound question about the nature of self-realization. It suggests that when everything becomes identified with the Self (Atman), there remains no distinction between the knower and the known. In this state of realization, there is no need to seek knowledge or identify any specific entity, as the essence of everything is realized to be the Self. This inquiry invites contemplation on the ultimate unity and identity of the self with all existence.

Freedom from Limitations:

- "The Self (Atman) is not affected by the actions of the body, just as space is not affected by the objects it contains." (Chandogya Upanishad 8.12.4)

  This quote illustrates the nature of the Self (Atman) as unaffected by the actions of the body. It draws a comparison with space, which remains unchanged regardless of the objects it contains. Similarly, the Self remains untouched by the experiences and actions of the body, indicating its inherent transcendence beyond the realm of worldly phenomena.

Atman is described as beyond all dualities and opposites, serving as the unchanging and eternal essence that remains unaffected by the fluctuations of the phenomenal world. Understanding the true nature of Atman involves transcending the limitations of dualistic thinking and recognizing the underlying unity of existence, leading to profound liberation and spiritual fulfillment.

Source of Happiness and Fulfillment:

Vedanta teaches that realizing the true nature of Atman brings about a state of profound happiness, peace, and fulfillment. By recognizing one's essential unity with Brahman, individuals are freed from the cycle of suffering and attain a state of inner harmony and contentment. It's like uncovering a hidden treasure that brings about a profound sense of joy and contentment.

Here's how the Upanishads portray this experience:

Realization of True Nature as the Source of Happiness:

- "When one realizes the Self (Atman) in every creature, one ceases to be indifferent to them. When one finds the Self (Atman) in oneself, one ceases to be indifferent to oneself." (Isha Upanishad 6)

  Here, the emphasis is on the transformative power of self-realization. The quote suggests that when one perceives the Self (Atman) in every living being, they develop a deep sense of connection and empathy towards all creatures, transcending indifference. Similarly, recognizing the Self within oneself leads to a profound self-awareness and self-compassion, eliminating indifference towards one's own being.

Freedom from Suffering:

- "The knower of Atman transcends sorrow, transcends death. By knowing Atman, one attains immortality." (Mundaka Upanishad 3.2.9)

  This excerpt highlights the transformative power of self-realization, inferring that the one who understands the nature of the Self (Atman) transcends sorrow and death, attaining a state of immortality. By gaining insight into the eternal essence within oneself, an individual transcends the limitations of mortality and achieves a timeless existence free of suffering.

Inner Harmony and Contentment:

- "When one knows Atman, which is hidden within the depths of one's being, then sorrow ceases, desires fade away, and the Self (Atman) is realized." (Chandogya Upanishad 7.1.3)

  The Chandogya Upanishad quote underscores the transformative impact of self-realization. By understanding the true nature of the Self (Atman), which resides within the depths of one's being, sorrow dissipates, desires diminish, and the individual attains realization of their true essence. This insight emphasizes the profound inner journey toward self-discovery and liberation from worldly afflictions.

State of Inner Fulfillment:

- "When all desires that dwell in the heart cease, then the mortal becomes immortal, and even in this very life, he reaches Brahman." (Chandogya Upanishad 8.1.6)

  This passage conveys the transformative journey toward spiritual liberation. It suggests that when all desires residing within the heart come to an end, the mortal transcends mortality and attains immortality. Through the cessation of desires, one achieves union with Brahman, the ultimate reality, even within the span of one's earthly existence. This insight emphasizes the profound significance of overcoming desires as a means to realize one's inherent divinity and attain liberation.

Vedanta teaches that realizing the true nature of Atman brings about a state of profound happiness, peace, and fulfillment. By recognizing one's essential unity with Brahman, individuals are freed from the cycle of suffering and attain a state of inner harmony and contentment, leading to profound liberation and spiritual fulfillment.

Path to Liberation:

Understanding and realizing the depths of Atman is central to the path of spiritual liberation in Vedanta. By recognizing the true nature of the self as Atman and understanding its identity with Brahman, individuals are liberated from the bondage of ignorance (avidya) and the cycle of birth and death (samsara), attaining ultimate freedom and union with the divine.

The sacred texts refer to this process in the following way:

Recognition of True Nature:

- "That art thou." (Chandogya Upanishad 6.8.7) This statement emphasizes the identity between the individual self (Atman) and the ultimate reality (Brahman), laying the foundation for the path to liberation.

  This statement encapsulates the profound truth of Vedantic philosophy, underscoring the intrinsic unity between the individual soul and the divine essence. It serves as a foundational principle for the spiritual journey toward liberation, highlighting the transformative potential of realizing one's true nature and recognizing the divine presence within oneself. By understanding the identity between the individual self and the ultimate reality, one embarks on the path of self-discovery and spiritual awakening, ultimately leading to liberation from worldly bondage.

Understanding Identity with Brahman:

- "The knower of Brahman becomes Brahman. He becomes all this, whatever exists." (Chandogya Upanishad 7.24.1)

  This declaration talks of the transformative power of spiritual knowledge, describing the profound shift that occurs when one attains true understanding of the ultimate reality. It emphasizes that the one who comprehends Brahman transcends the limitations of individual identity and merges with the universal essence. This insight underscores the transformative nature of spiritual wisdom, suggesting that genuine knowledge leads to a profound shift in consciousness, allowing the individual to recognize their intrinsic oneness with the entire existence. In essence, it unveils the boundless potential for self-realization and unity with the cosmic reality.

Freedom from Ignorance and Suffering:

- "The self (Atman), smaller than small, greater than great, is hidden in the heart of every creature. A man who is free from desires and free from grief sees the majesty of the self (Atman) through tranquility of mind and senses." (Katha Upanishad 1.2.20)

  Described as "smaller than small, greater than great," the Atman is depicted here as an omnipresent essence, existing within the heart of every being, regardless of size or stature. The quote highlights that the realization of the self (Atman) is not contingent upon external factors but is attainable through inner tranquility and freedom from desires and sorrows. By quieting the fluctuations of the mind and senses, one gains the clarity to perceive the inherent magnificence of the self (Atman).

Attainment of Ultimate Freedom:

- "The wise, having realized the Self (Atman) as dwelling within all beings, becomes compassionate towards all creatures. What delusion or sorrow can there be for the one who beholds unity everywhere?" (Isavasya Upanishad 6)

  The above quote underscores that the realization of the self (Atman) within all beings leads to a state of absolute freedom. When one comprehends that the same divine essence dwells within every individual, it fosters a sense of interconnectedness and compassion. This understanding transcends the illusions

> of separation and ego, liberating the individual from the constraints of selfish desires and worldly attachments. By beholding unity everywhere, one attains a state of inner peace and contentment, free from the sufferings of delusion and sorrow. In this way, recognizing the presence of the self in all beings becomes a pathway to absolute freedom and liberation.

Embarking on the journey of exploring Atman is akin to diving into the depths of one's own existence, peeling away the layers of illusion to uncover the jewel of ultimate truth. It's a quest that takes us beyond the surface-level identities we often cling to and invites us to discover the timeless essence that animates our very being. As we delve deeper into the inner realms of consciousness, we come to realize that Atman is not just a concept or an abstract idea but the very core of our existence — the spark of divinity that resides within us all.

This journey of self-discovery is not merely a process of intellectual inquiry but a profound spiritual odyssey that touches the essence of our soul. It's about awakening to the eternal truths that have been whispered by the sages across the ages, recognizing our oneness with the cosmic whole, and tapping into the infinite reservoir of love and wisdom that flows from the source of all creation.

Through this exploration, we come to understand that Atman is not separate from Brahman but an indivisible part of the universal consciousness. It's the recognition that our individual selves are but reflections of the divine, mirroring the infinite beauty and boundless potential of the cosmos. And as we unravel the mysteries of Atman, we unlock the door to spiritual liberation — the freedom from the shackles of ignorance and the realization of our true nature as divine beings.

Ultimately, the journey into Atman is a sacred pilgrimage of the soul — a quest for self-realization and enlightenment that leads us to the pinnacle of human experience. It's a journey of transformation, transcendence, and profound inner awakening — a journey that beckons us to dive deep into the ocean of consciousness and emerge reborn, radiant, and liberated.

And most off, please remember – it's the journey where the greatest of bliss lies, a journey that is itself the destination!

In the next part of this book, we will delve into the diverse paths to transcendence within the rich tradition of Vedanta. Join me as we explore the wisdom and depth of these paths.

Touching the Atman:

During a phase in my life when I was exploring different spiritual healing traditions, I decided to travel to an island in Indonesia known for its amazing healers. I had heard stories that the rational part of my mind found very difficult to believe so, obviously, I decided to visit. ☺

On arriving in this particular village, I witnessed a huge crowd and a long queue leading into a hut, where the healer would perform his rites. It seemed to be straight out of a movie set. After asking around a bit, I managed to find the elderly gentleman I had been referred to, who was supposed to get me access to see the Chief Healer. After a couple of hours, I was invited by the healer, who, surprisingly, spoke flawless English. I later learnt that he had worked in the UK for a bit as an engineer. He asked me the purpose of my visit and when I shared that I was studying healing systems, he looked at me and said, *"You don't seem to have healing as your duty but I can see you have the soul of a teacher, so I will show you. Come back at 5 pm."* I went back at 5 pm and saw some amazing healing rituals and results, none of which I am allowed to share here. Let's just say they are some of the most amazing things I have ever seen and even more difficult for me to believe than if I saw a person walking on water. Just incredibly humbling.

Somewhere between 10 pm and midnight, they asked me if I wanted to receive a healing and if I had a specific request. I told them I didn't, following which I received a healing from the Chief Healer. I can't go into the specifics but suddenly, five minutes into the experience, I had a searing pain in my head, as if it was going to split into two, followed by a deep burning sensation down my spine. The pain was quite intense, literally causing tears to flow and I did my best not to cry.

It is probably appropriate to clarify that I did not ingest/drink or smoke anything before the pain started. The rituals merely included some energy work and chanting. This pain lasted for about 20 excruciating minutes before disappearing as quickly as it had appeared. I did feel nausea for another hour or so, then it eventually passed. Left feeling totally exhausted, I was told by the healer to go and rest and see him in the morning. I slept very soundly for a long time – almost 10 hours, which is not normal for me.

The next morning I woke up late feeling very light and clean – it's a feeling very hard to put into words. Later in the day I met with the Chief Healer and asked him what had happened. He told me I had received a powerful healing and I should feel light and clean. I told him that I did, but what about the pain? He replied with a mysterious smile, *"Oh, that. It was just you feeling the touch of your soul cleansing your body. But it was just a touch, now you are ready to awaken your soul connection more."*

I was stunned to hear it and quickly asked, *"What does that mean? What happens next?"*

He smiled and replied, *"Always the impatient seeker you are. Next is what your soul desires, and only that will happen, what is meant to happen and must happen. Go now, find your life."*

I then said goodbye and returned back to my accommodation before making it back to the city enroute to my home.

Over time, I realized the significance of that day although it's taken me almost a decade to integrate that experience. I am not quite sure what happened and what it meant, but over the years I have begun to realize that understanding it is not as important as being guided by it and learning from it. I have since then become a more *patient seeker with a greater degree of surrender to my soul.*

"The teachings of Vedanta and Buddhism are the essence of the wisdom of the East, and their understanding can lead to the highest spiritual realization."

– Hermann Hesse

PART II:

PATHS TO TRANSCENDENCE

What does the term 'transcendence' mean to you? And, why (if at all) is it significant to experience transcendence on the way to enlightenment?

From a spiritual standpoint, transcendence usually refers to the experience of moving beyond the ordinary, mundane aspects of existence and accessing higher states of consciousness or awareness. It involves surpassing the limitations of the ego, the material world, and conventional modes of perception to connect with something greater than oneself.

Spiritual transcendence can lead to a sense of unity with the divine, the universe, or the ultimate reality, bringing feelings of profound peace, bliss, and interconnectedness.

There are various spiritual practices within Vedanta that can aid transcendence, such as meditation, prayer, self-inquiry, or acts of selfless service. It is, in effect, a transformative journey towards spiritual awakening and enlightenment.

If transcendence is moving beyond the limitations of ordinary perception and the egoic mind, enlightenment entails transcending the illusory boundaries of the individual self and recognizing the interconnectedness of all existence.

When we break free of the confines of the ego and the material world, we can gain insight into the impermanent nature of phenomena. This brings about a profound shift in perception, leading to a deeper understanding of the nature of reality and one's place within it.

Moreover, transcendence opens the door to experiencing higher states of consciousness characterized by peace, joy, and unity with the divine. It allows us to tap into our innate spiritual potential and access the boundless wisdom and love that lie at the core of our being.

Ultimately, transcendence serves as a gateway to enlightenment by helping us realize the true nature of existence and experience the ultimate reality beyond the confines of our egoic mind.

According to Vedanta, there are various paths or approaches to transcendence.

These paths, known as yogas, provide different avenues for spiritual seekers to attain liberation (moksha) and understand their essential unity with the ultimate reality, Brahman.

Each yoga caters to the unique temperaments and inclinations of individuals, providing various methods and practices for spiritual growth and self-realization.

- **Jnana Yoga,** the path of knowledge, involves intellectual inquiry and discernment to understand the true nature of reality and the self.
- **Bhakti Yoga**, the path of devotion, emphasizes cultivating a loving relationship with the divine through prayer, worship, and surrender.
- **Karma Yoga,** the path of selfless action, encourages performing duties without attachment to the results, dedicating actions to the divine and serving humanity.
- **Raja Yoga,** the path of meditation and mental control, focuses on practices such as

concentration, meditation, and breath control to quiet the mind and attain union with the divine.

Together, these paths offer comprehensive approaches to spiritual growth and eventual liberation, guiding seekers towards the realization of their essential unity with Brahman and the attainment of eternal peace and bliss.

Let's take a look at the different paths of Vedanta in more detail to understand how we can transcend our present reality on the way to enlightenment.

"Quantum concepts will now no longer appear ridiculous to humans who've studied Vedanta."

– Werner Heisenberg

CHAPTER 7

INTELLECT ILLUMINATION: JNANA YOGA'S PURSUIT OF WISDOM

I want to introduce you to Jnana Yoga, known as the Path of Knowledge; a quest for spiritual enlightenment that is intertwined with the fervent pursuit of intellectual understanding. This profound path beckons seekers to dive deep into the sacred scriptures, engaging in rigorous philosophical inquiry and contemplation to unravel the mysteries of reality, the self, and the transcendent Brahman.

In the realm of Jnana Yoga, the ultimate goal is to pierce through the veil of *ignorance (avidya)* and awaken to the profound realization of non-duality within existence. Here, the ancient teachings of Vedanta, especially the profound philosophy of Advaita Vedanta, illuminate the path, urging practitioners to embark on a journey of self-discovery through relentless self-inquiry and discernment, distinguishing the eternal truth of Brahman from the fleeting illusions of the material world.

This profound path traces its roots deep into the annals of Indian spiritual tradition, where sages and scholars alike have delved into the depths of wisdom to unravel the mysteries of existence. Throughout history, luminaries such as Adi Shankaracharya, the great 8th-century philosopher and exponent of Advaita Vedanta, have left an indelible mark on the tradition of Jnana Yoga. Shankaracharya's profound commentaries on the Upanishads and Bhagavad Gita continue to inspire seekers on the path of self-realization even today.

In contemporary times, people like Ramana Maharshi, the revered sage of Arunachala, stand as beacons of wisdom in the realm of Jnana Yoga. Ramana Maharshi's simple yet profound teachings on self-inquiry, encapsulated in his famous directive "Who am I?" have guided countless seekers towards the realization of the true nature of the Self.

Another towering figure in the modern era is Swami Vivekananda, the eloquent disciple of Sri Ramakrishna Paramahamsa. Vivekananda's stirring speeches at the Parliament of the World's Religions in 1893 brought the ancient wisdom of Jnana Yoga to the global stage, igniting the flames of spiritual inquiry and fostering a deeper understanding of Vedanta in the Western world.

One anecdote that encapsulates the essence of Jnana Yoga is the famous encounter between the sage Ashtavakra and King Janaka we find in the Ashtavakra Gita, a revered scripture of Advaita Vedanta. In this timeless dialogue, Ashtavakra challenges King Janaka's understanding of reality, leading him to the profound realization of his true nature beyond the confines of worldly identities. This scripture is considered a classic text of Jnana Yoga and is highly regarded for its teachings on non-duality and self-realization.

By embarking on this transformative odyssey into Jhana Yoga, we will discover how the wisdom of the ages converges with contemporary insights, guiding us on the timeless quest for self-realization and liberation.

Emphasis on Pursuit of Spiritual Wisdom:

In the labyrinth of human existence, where the depths of the soul meet the vast expanse of the cosmos, Jnana Yoga unveils the elusive treasure known as spiritual wisdom. Far beyond the realm of mere intellectual understanding, spiritual wisdom in Jnana Yoga is a beacon of light that illuminates the path to ultimate truth and liberation.

At its core, spiritual wisdom is not merely about accumulating knowledge or reciting philosophical doctrines; it is a sacred journey of inner exploration, a quest to unravel the mysteries of existence and uncover the eternal essence that lies dormant within every being.

In the tradition of Jnana Yoga, spiritual wisdom is the art of discernment, the ability to pierce through the veil of illusion and perceive the underlying unity of all creation. It is the realization that beyond the transient phenomena of the material world lies an eternal reality, the divine essence known as Brahman, which permeates every atom of existence.

But spiritual wisdom is not confined to lofty concepts or abstract philosophies; it is a living, breathing force that permeates every aspect of life. It is the wisdom to discern truth from falsehood, love from illusion, and liberation from bondage. It is the clarity of vision that enables one to navigate the complexities of life with grace and equanimity, undeterred by the storms of circumstance.

Spiritual wisdom, then, is not a destination to be reached but a journey to be embraced — a journey of self-discovery and self-realization. It is the key that unlocks the gates of liberation, the guiding star that leads seekers from the darkness of ignorance to the dawn of enlightenment. It is a timeless treasure waiting to be discovered, beckoning all who are willing to embark on the sacred quest for truth and transcendence.

- Jnana Yoga places a strong emphasis on seeking spiritual wisdom as the primary means to attain liberation. This wisdom is not merely intellectual knowledge but an experiential understanding of the ultimate truth.
- Example: You may study the Upanishads, Vedanta Sutras, or Bhagavad Gita to gain insights into the nature of reality and the self. Since these texts are very rich and complex, you may need to study them under the guidance of a qualified teacher (guru) to understand the spiritual principles and their application in daily life.
- Source: *"He who knows Brahman attains the highest. Having known that Brahman, he becomes the highest among the knowers of Brahman." (Mundaka Upanishad 1.3.9)*

  This quote underscores the importance of knowing Brahman, the ultimate reality. It suggests that those who attain this knowledge achieve the highest spiritual realization. Furthermore, it implies that upon realizing Brahman, the seeker ascends to the pinnacle of wisdom, surpassing even the most learned scholars. In essence, the quote highlights the transformative power of spiritual enlightenment in attaining ultimate truth and understanding.

Rigorous Study of Scriptures and Philosophical Inquiry:

In Jnana Yoga, the pursuit of spiritual wisdom is not a passive endeavor but a thrilling odyssey that unfolds through rigorous study of scriptures and profound philosophical inquiry. Here, amidst the ancient texts and timeless teachings, seekers embark on a quest that transcends the boundaries of the mind and delves into the depths of the soul.

At the heart of Jnana Yoga lies the recognition that true knowledge is not handed down on a silver platter but earned through diligent effort and unwavering dedication. The rigorous study of scriptures serves as the compass that guides seekers through the labyrinth of existence, offering profound insights into the nature of reality and the path to spiritual liberation.

Through the scriptures, seekers encounter the wisdom of sages and saints who have traversed the path before them, illuminating the way with their timeless teachings and profound insights. Each verse, each sutra, becomes a stepping stone on the journey of self-discovery, inviting seekers to unravel the mysteries of existence and awaken to the eternal truths that lie dormant within.

But the study of scriptures is merely the beginning — a launching pad for the soul's ascent into the realm of higher knowledge. In Jnana Yoga, philosophical inquiry is the engine that propels seekers beyond the realm of mere intellectual understanding and into the domain of direct experience.

Through philosophical inquiry, seekers engage in a profound dialogue with the fundamental questions of existence — Who am I? What is the nature of reality? What is the purpose of life? These questions become gateways to deeper understanding, inviting seekers to peel back the layers of illusion and uncover the eternal essence that lies at the heart of all creation.

In essence, the importance of rigorous study of scriptures and philosophical inquiry in Jnana Yoga cannot be overstated. They are the twin pillars that support the edifice of spiritual wisdom, offering seekers the tools they need to navigate the complexities of existence and awaken to the divine truth that dwells within. In the hallowed halls of Jnana Yoga, the pursuit of knowledge is not merely an academic exercise but a sacred quest — a journey of self-discovery and enlightenment that leads seekers from darkness to light, from ignorance to wisdom, and from bondage to liberation.

- Jnana Yoga involves a rigorous study of sacred texts and engaging in philosophical inquiry to contemplate the deeper meanings of spiritual teachings. This inquiry helps the seeker to develop clarity of thought and discernment.
- Example: You may study verses from the Upanishads that discuss the nature of Brahman, the self, and the relationship between the two. You can also engage in discussions with fellow seekers and teachers to explore different interpretations and deepen your understanding of the teachings.

- Source: *"Let a man take to the study of the Vedas (scriptures) and also to meditation (contemplation)."* (Chandogya Upanishad 7.1.1)

  Here, we can understand the importance of both studying the Vedas and engaging in meditation. It encourages individuals to immerse themselves in the wisdom of the scriptures while also cultivating a practice of contemplation and introspection. In essence, the quote highlights the complementary nature of intellectual inquiry and inner reflection in the pursuit of spiritual knowledge and realization.

Contemplation on the Nature of Reality, Self, and Brahman:

In Jnana Yoga, the exploration of the nature of reality, self, and Brahman unfolds as a spellbinding saga — a cosmic dance of inquiry that transcends the boundaries of time and space, leading seekers on a profound quest for truth and enlightenment.

Jnana Yoga recognizes that the nature of reality is not merely a philosophical concept to be debated but a profound mystery waiting to be unraveled. Here, amidst the ephemeral mist of the material world, seekers delve deep into the fabric of existence, peeling back the layers of illusion to uncover the eternal essence that lies at its core.

Central to this exploration is the inquiry into the nature of the self — the eternal spark of consciousness that animates all of creation. In Jnana Yoga, the self is not merely an individual entity confined to the limitations of the physical body but a boundless ocean of consciousness that transcends all boundaries. Through introspection and self-inquiry, seekers come to realize their true identity as the eternal Self (Atman), which is indistinguishable from the ultimate reality known as Brahman.

Ah, Brahman — the divine essence that permeates the cosmos like a silent melody echoing through the halls of eternity. In Jnana Yoga, Brahman is not merely a distant deity to be worshiped but the very fabric of existence — the substratum upon which the entire universe rests. Through contemplation and meditation, seekers come to recognize Brahman as the source and sustainer of all creation — the eternal wellspring of life, consciousness, and bliss.

But the importance of understanding the nature of reality, self, and Brahman in Jnana Yoga extends far beyond mere intellectual curiosity. It is a quest for self-realization, a journey of awakening that leads seekers from the shadows of ignorance to the radiant light of enlightenment. Through the exploration of these timeless truths, seekers come to transcend the limitations of the ego and merge with the infinite expanse of divine consciousness, experiencing the ultimate freedom and liberation that lies at the heart of existence.

In essence, the exploration of the nature of reality, self, and Brahman in Jnana Yoga is a sacred pilgrimage — a voyage of discovery that leads seekers from the shores of illusion to the shores of truth. It is a journey that invites all to unlock the mysteries of existence and awaken to the eternal truths that lie dormant within.

- Contemplation is a crucial aspect of Jnana Yoga, as it allows the seeker to introspect and reflect on the nature of reality and the self. Through contemplation, the seeker aims to go beyond intellectual understanding and directly experience the truths expounded in the scriptures.
- Example: You may practice meditation on verses that describe the nature of Brahman as beyond attributes and distinctions. By dwelling on these teachings with focused attention, you will gradually gain deeper insights into the nature of ultimate reality.
- Source: *"That which is invisible, imperishable, unknowable, ungraspable, without source, without attributes, without stain, that through which the breaths are breathed forth — That alone is the Self (Atman), and That is to be realized."* (Mundaka Upanishad 2.1.2)

  The nature of the Self (Atman) is shown to be beyond the grasp of the senses and conceptual understanding. The quote describes the Self as invisible, imperishable, and unknowable, transcending all attributes and origins. The verse directs seekers to realize the Self, which is the essence underlying all existence and the source of life itself. In essence, the quote underscores the importance of realizing the profound truth of the Self, which lies beyond the realm of perception and intellectual comprehension.

Transcending Ignorance (Avidya) and Realizing Non-Dual Nature:

The journey towards enlightenment in Jnana Yoga unfolds as a mesmerizing tale — a cosmic odyssey that traverses the depths of ignorance and ascends to the dizzying heights of non-dual realization. At its core lies the imperative of transcending ignorance (Avidya) and realizing the non-dual nature of existence — a quest that beckons seekers to pierce through the veils of illusion and awaken to the boundless truth that lies beyond.

Ignorance, like a dense fog shrouding the soul, veils the luminous essence of reality and obscures the eternal truths that lie dormant within. In Jnana Yoga, ignorance is not merely a lack of knowledge but a fundamental misunderstanding of the nature of reality — a veil that blinds seekers to the radiant light of divine wisdom.

But the journey of Jnana Yoga is one of illumination — a sacred pilgrimage that leads seekers from the shadows of ignorance to the radiant shores of enlightenment. Through rigorous study, contemplation, and self-inquiry, seekers peel back the layers of illusion, unraveling the mysteries of existence, and transcending the limitations of the ego.

Central to this journey is the realization of the non-dual nature of existence — a profound awakening that lies at the heart of Jnana Yoga. Here, amidst the tapestry of creation, seekers come to recognize the inherent unity that underlies all of existence — the seamless fabric of consciousness that weaves through the cosmos.

In this realization, the boundaries between self and other, subject and object, dissolve like dewdrops in the morning sun, revealing the eternal truth of non-duality. It is a realization that transcends the limitations of the mind and opens the floodgates of divine wisdom, flooding the soul with boundless love, compassion, and understanding.

Essentially, the journey of Jnana Yoga is one of awakening, of transcending ignorance, and realizing the radiant truth of non-duality — a journey that invites all to walk on the path of self-discovery and enlightenment.

- Jnana Yoga seeks to transcend the veil of ignorance (avidya) that obscures the true nature of reality. By piercing through this ignorance, the seeker realizes the non-dual nature of existence, where the individual self (Atman) is identified with the ultimate reality (Brahman).
- Example: Through contemplation and self-inquiry, you may come to recognize that the sense of individuality and separateness is an illusion. You realize that the true self (Atman) is not limited by time, space, or causality and is identical with Brahman, the ultimate reality.
- Source: *"When the five sheaths of the Self (Atman) are pierced, the knower sees the Pure Self within, realizes Brahman, and attains immortality."* (Taittiriya Upanishad 2.1.1)

  This quote from the Taittiriya Upanishad illuminates the journey of self-realization and ultimate enlightenment. It describes the process by which the seeker penetrates through the five layers or sheaths that veil the true nature of the Self (Atman). Once these sheaths are transcended, the knower gains the vision of the Pure Self within, recognizes the divine essence of Brahman, and achieves immortality. In essence, the quote highlights the transformative journey of piercing through layers of illusion to realize the eternal truth of the Self and attain liberation from the cycle of birth and death.

Importance of Self-Inquiry and Discrimination:

The path to enlightenment in Jnana Yoga unfolds as a sacred voyage of self-discovery that beckons seekers to plunge into the depths of their own being and unravel the mysteries of existence. At its heart lies the transformative practices of self-inquiry and discrimination — tools that illuminate the path to truth and liberation with the brilliance of a thousand suns.

Self-inquiry, like a compass guiding the soul through the labyrinth of existence, invites seekers to turn their gaze inward and explore the depths of their own consciousness. It is a profound quest to peel back the layers of illusion and uncover the radiant essence of the Self (Atman) that lies dormant within. Through introspection and contemplation, seekers come to question the very nature of their existence, probing the depths of their being with unyielding curiosity and relentless determination.

But self-inquiry is not merely an intellectual exercise; it is a journey of self-discovery that leads seekers to the core of their being. Here, amidst the tapestry of thoughts, emotions, and sensations, seekers come to realize their true identity as the eternal Self, transcending the limitations of the ego and merging with the boundless expanse of divine consciousness.

Central to the practice of self-inquiry is discrimination — the razor-sharp sword that cuts through the illusions of the mind and discerns truth from falsehood. In Jnana Yoga, discrimination is not about dividing the world into categories of good and bad, right and wrong; it is about piercing through the veils of ignorance and realizing the ultimate truth that lies beyond.

Through discrimination, seekers come to recognize the transient nature of the material world and discern the eternal essence that underlies all of creation. It is a process of separating the real from the unreal, the eternal from the transient, and awakening to the boundless truth of non-duality that pervades all of existence.

But the importance of self-inquiry and discrimination in Jnana Yoga extends far beyond mere philosophical speculation. It is a quest for liberation — a journey of self-discovery that leads seekers from the shadows of ignorance to the radiant shores of enlightenment. In the depths of self-inquiry and discrimination, seekers come to realize the ultimate truth — that they are not mere mortals bound by the chains of ego and desire, but divine beings of infinite potential and boundless freedom.

In essence, the practices of self-inquiry and discrimination in Jnana Yoga are sacred rites that lead seekers from darkness to light, from ignorance to wisdom, and from bondage to liberation. It is a journey that promises not only the realization of truth but the experience of ultimate freedom and bliss.

- Self-inquiry *(Atma-vichara)* and discrimination (Viveka) play crucial roles in Jnana Yoga. Through self-inquiry, the seeker investigates the nature of the self and discerns between the eternal self (Atman) and the transient aspects of existence. Discrimination helps the seeker distinguish between the real (Brahman) and the unreal (world of appearances).
- Example: You may ask yourself questions like "Who am I?" to probe the nature of the self beyond the body, mind, and ego. Through discriminative inquiry, you learn to differentiate between the changing phenomena of the world and the unchanging reality of Brahman.
- Source: *"He who knows the Self (Atman) as 'I am Brahman' becomes all this universe."* (Chandogya Upanishad 3.14.1)

  The above quote enshrines the profound realization of the identity between the individual self (Atman) and the universal essence (Brahman). It declares that one who recognizes the Self as identical to Brahman realizes the interconnectedness of all existence. By understanding oneself as Brahman, the essence underlying the entire universe, the individual transcends the boundaries of individuality and becomes one with the cosmic fabric of creation. In essence, the quote emphasizes the transformative power of realizing the unity of the self with the universe, leading to a profound sense of oneness and spiritual liberation.

In essence, Jnana Yoga involves a systematic approach to spiritual inquiry and understanding. This ancient path beckons seekers with the promise of unraveling the mysteries of existence and transcending the cycle of birth and death. Here, the quest for spiritual enlightenment is not a mere journey but a profound odyssey of self-discovery and liberation that leads to the ultimate truth. Through a meticulous blend of study, contemplation, and relentless self-inquiry, practitioners embark on a journey that unveils the hidden dimensions of reality.

Jnana Yoga is a path of awakening, where the mind becomes a powerful tool in the quest for transcendence; a path that not only enlightens the mind but also liberates the soul!

Is that all you know?

During an early phase of my study of energy medicine, I was actively being personally trained by a teacher who was truly a master healer. I am so very grateful that I got to travel with him for a long time, supporting him in all the events and healing camps he went to. On one such journey, we ended up in London, where a key organizer I will call Mani had organized a workshop. Once the workshop was over, she told us that her uncle who was a very successful doctor had invited us home for tea.

As we were driving there, she received a call from her uncle and spoke to him in *Gujarati,* which is an Indian language. While I could not fully understand what was discussed, I figured that they had also invited a well known psychic to meet with us that afternoon. I later learned that the psychic was someone who had suffered from a stroke, lost partial movement of his limbs, but developed a very strong psychic faculty. He was able to accurately predict the future for others by "channeling masters." I was quite curious about meeting him.

We arrived at the doctor's house, made some small talk, and I quickly realized that the whole group were devotees of Sai Baba (an Indian spiritual teacher/guru). After my teacher had delivered a small talk and led a brief meditation, the conversation shifted to "channeling" and "psychic faculties" in general. Everyone was extremely interested in the psychic, who I then realized was quite a celebrity in the group.

During the conversation, my teacher had remarked that channeling was not safe as you never know who you are channeling and how it may affect you. He further said something to the effect of *channeling not being a big deal*. Of course, he said it very politely and to make a point but it drew the immediate attention of the psychic, who challenged my teacher by asking him if he knew how to intentionally select and channel "clean beings."

My teacher smiled, replying that he did, and asked the psychic to close his eyes. He then made some energy healing gestures (which I recognized but I am not at liberty to disclose, hence, I am leaving it purposefully vague). Suddenly, the whole demeanor of the psychic changed. His previously shaking hands, due to the early stages of Parkinson's and his stooped over demeanor seems to disappear as he sat up straight, looking ahead with steely eyes. Next, he began speaking in a booming voice in *Hindi*, addressing my teacher and asking him, *"Why did you leave India? We had so much work to do together. What are you doing here?"* I knew my teacher didn't understand Hindi so I quickly started taking notes. Then, the psychic turned to me, addressing me in English. *"You think you are so smart but you know only IT. Baba knows all languages. I am talking directly to him,"* and pointed the finger at my teacher.

I was taken aback! Nobody from that gathering knew I was an IT professional. More words were spoken, which I am not allowed to disclose, and this "channeling" stopped as suddenly as it had started. The psychic went back to being himself with no recollection of what had transpired. There was a stunned silence in the room. Eventually the doctor spoke up, saying, *"This was the voice of Baba. I remember it from the time I used to treat him."* (Baba here refers to Sai Baba). Everyone was shocked to hear that and all eyes turned to my teacher. He just smiled and said, *"See? I told you it isn't a big deal if you know how."* The next few minutes were a blur before we quickly got up and left.

When in the car alone, I asked my teacher how it was that he got the guy to channel Sai Baba. He replied, *"It's like how you would call V.A. and ask him to leave a voicemail* (V.A. was my business partner back then). We passed the rest of the journey in complete silence and soon arrived at our hotel. Before parting for the night, my teacher said, *"Leo, your journey has been of Jnana Yoga. It's about time you started exploring what lies beyond that. There are many paths with their own learnings."* This marked a change in my learning style forever, I began to seek out and work on synthesis. And that remains my dutiful pursuit till today!

"Bhakti Yoga is the path of love and devotion. It teaches that through devotion to God, one can attain the highest spiritual realization. Love is the highest of all spiritual practices."

– Christopher Isherwood

CHAPTER 8

HEART'S QUEST: BHAKTI YOGA WITHOUT A PERSONAL GOD

Bhakti Yoga is an enchanting realm where the fervor of devotion ignites the soul's journey towards the divine. Traditionally seen as the path of unwavering adoration towards a personal deity, Bhakti Yoga unveils a deeper, more profound facet when practiced beyond the confines of anthropomorphic perceptions of God.

Here, the heart's devotion transcends mere forms and images, delving into the boundless expanse of the divine essence itself. It's a shift from worshiping a particular deity to revering the timeless force that breathes life into every atom of creation.

In the practice of Bhakti Yoga, the practitioner's spirit becomes a vessel overflowing with profound love and reverence for the ultimate reality — the very source from which all existence springs forth. It's a recognition of the divine essence as the pulsating heartbeat of the cosmos, the unseen thread that weaves together the fabric of the universe.

With each heartfelt prayer and every soul-stirring chant, the practitioner dives deeper into the ocean of divine love, merging with the cosmic symphony that reverberates throughout creation. It's a journey of surrender, where the ego dissolves in the boundless melodies of divine grace, and the soul dances in ecstatic union with the infinite.

Throughout history, Bhakti Yoga has been championed by luminaries whose devotion illuminated the path for countless seekers. Figures like Mirabai, the 16th-century mystic poetess, epitomized the essence of Bhakti Yoga through her impassioned love for Lord Krishna, composing soul-stirring verses that continue to resonate with devotees today.

In contemporary times, the teachings of Bhakti Yoga have been revitalized by spiritual leaders like Swami Sivananda Saraswati, founder of the Divine Life Society, whose emphasis on love, devotion, and self-surrender inspired millions worldwide to tread the path of Bhakti with unwavering faith.

Anecdotes abound in the Bhakti tradition, illustrating the transformative power of devotion. One such example revolves around the life of Tulsidas, the 16th-century saint-poet revered for his epic poem, the Ramcharitmanas. Legend has it that Tulsidas' unwavering devotion to Lord Rama moved the very elements of nature, as depicted in stories of miracles attributed to him, such as the episode where he made the river Yamuna part to allow him to cross.

But Bhakti Yoga isn't confined to ancient tales; its timeless essence continues to inspire contemporary practitioners. In the bustling streets of modern India, one can witness the fervent devotion of millions who flock to temples, chanting the names of their beloved deities with hearts aflame with love and reverence.

Let's take a look at this path of divine connection that invites seekers to lose themselves in the divine embrace of unconditional love.

Cultivating a Loving Relationship:

In Bhakti Yoga without a personal God, the emphasis is on cultivating a loving relationship with the divine essence, which is perceived as an all-encompassing presence that permeates every aspect of existence. As a practitioner, you would approach this divine essence with reverence, love, and devotion, seeking to establish a heartfelt connection through various spiritual practices.

The cultivation of a loving relationship begins with a deep sense of awe and wonder towards the magnificence of the cosmos and the inherent divinity present in every being. This reverence serves as the foundation for developing a heartfelt connection with the divine essence. You would view every aspect of existence as a manifestation of the divine, seeing the beauty and perfection inherent in all creation.

Love becomes the driving force behind your spiritual practice, guiding you to express gratitude, compassion, and kindness towards all beings. You would seek to nurture an intimate and personal relationship with the divine essence, understanding that it is not separate from you but rather the very essence of your being.

Various spiritual practices serve as pathways for establishing and deepening this loving relationship. Meditation, contemplation, chanting of sacred mantras, and acts of selfless service (seva) are all means through which you can commune with the divine essence and experience its presence in your life.

Through these practices, you gradually dissolve the barriers of separation between yourself and the divine, realizing the interconnectedness and unity of all existence. This realization deepens your love and devotion, fostering a sense of oneness with the divine essence that permeates every aspect of your being.

Below are some of the source texts, with explanations to help you gain an insight into their wisdom:

- "The wise, who perceive the underlying unity of all beings and see the same Self in everyone, cultivate love and compassion towards all." (Isavasya Upanishad 6)

  This text is a reflection of the profound wisdom found in the Isavasya that encapsulates the essence of spiritual realization. According to this verse, those who are wise understand that there is a fundamental unity underlying the diversity of life and perceive the same divine essence (the Self) in every individual. As a result, they naturally cultivate love and compassion towards all beings, transcending boundaries of ego and self-interest.

- "He who sees all beings in the Self (Atman), and the Self in all beings, feels no hatred by virtue of that realization. For him, who is aware that all beings are the Self, there is nothing that he does not love." (Isavasya Upanishad 6)

  This additional quote from the Isavasya Upanishad further emphasizes the profound realization of unity and interconnectedness in Bhakti Yoga. It highlights the perspective of the wise who perceive the divine essence (the Self) pervading all beings. By recognizing the Self in every individual and seeing all beings as manifestations of the same divine essence, the practitioner transcends the barriers of ego and selfishness. As a result, they experience a deep sense of love and compassion towards all of creation.

Ultimately, cultivating a loving relationship in Bhakti Yoga without a personal God leads to a profound sense of fulfillment, inner peace, and joy. It transforms your perception of reality, guiding you to see the divine presence in every moment and every experience, and inspiring you to live a life filled with love, compassion, and divine grace.

Prayer:

In Bhakti Yoga without a personal God, prayer serves as a profound means of expressing devotion and surrendering to the divine essence that pervades all existence. Unlike traditional forms of prayer directed towards a specific deity with anthropomorphic qualities, prayer in this context is directed towards the abstract and universal divine essence that transcends all forms and attributes.

Prayer becomes a sacred dialogue between the practitioner and the divine essence, a heartfelt expression of reverence, love, and surrender. Through prayer, the practitioner communicates their deepest aspirations, desires, fears, and gratitude, laying bare their innermost thoughts and feelings before the divine presence.

One of the key aspects of prayer is the act of surrendering to the divine will. The practitioner acknowledges their limitations and dependencies, recognizing the omnipotence and omniscience of the divine essence. In surrendering, they relinquish their egoic desires and attachments, trusting in the wisdom and benevolence of the divine plan.

It also fosters a sense of intimacy and communion with the divine essence. As the practitioner opens their heart in prayer, they experience a profound connection with the divine presence, feeling enveloped in love, grace, and divine guidance. This intimate communion deepens over time through consistent practice and sincere devotion.

There are various forms of prayer that can be practiced, including spoken words, silent contemplation, devotional songs (bhajans), and recitation of sacred mantras. Each form of prayer carries its own unique potency, serving as a vehicle for expressing devotion and establishing a direct connection with the divine essence.

Through prayer, an individual not only communicates their own needs and desires but also expresses gratitude for the countless blessings bestowed upon them. Gratitude becomes a powerful spiritual practice, fostering a deep sense of appreciation and humility before the divine presence.

We can gleam more insights into the power of prayer in the quotations below:

- "Whatever desires people bring to me in prayer, I fulfill them all. People, therefore, worship me with various desires in their hearts." (Brihadaranyaka Upanishad 4.4.22)

  This text from the Brihadaranyaka Upanishad emphasizes the responsive nature of the divine towards the desires expressed in prayer. It suggests that the divine fulfills all desires brought forth by devotees in prayer, highlighting the significance of prayer as a means of seeking fulfillment and expressing devotion in Bhakti Yoga.

- "Offer your mind to Me, be devoted to Me, worship Me, and bow down to Me. Doing so, you will come to Me alone, I truly promise you, for you are exceptionally dear to Me." (Bhagavad Gita 9.34)

  In this verse from the Bhagavad Gita, Lord Krishna speaks to Arjuna, emphasizing the importance of devotion and surrender in Bhakti Yoga. Here, Krishna assures that those who offer their minds to Him, remain devoted, worship Him, and bow down in reverence will undoubtedly attain union with Him. This verse underscores the significance of prayer and devotion as fundamental aspects of Bhakti Yoga, highlighting the intimate relationship between the devotee and the divine.

- "I am seated in everyone's heart, and from Me come remembrance, knowledge, and forgetfulness. By all the Vedas, I am to be known; indeed, I am the compiler of Vedanta, and I am the knower of the Vedas." (Bhagavad Gita 15.15)

  In this verse, Lord Krishna elucidates His omnipresence and omniscience. He declares that He resides in the heart of every being, facilitating remembrance, knowledge, and forgetfulness.

> Furthermore, Krishna asserts that He is the ultimate subject of knowledge expounded in the Vedas. This verse underscores the importance of prayer in Bhakti Yoga by acknowledging the divine presence within every individual and highlighting the role of the divine in bestowing knowledge and understanding.

Overall, prayer in Bhakti Yoga without a personal God is a sacred and transformative practice that nourishes the soul, deepens devotion, and fosters a sense of intimacy and communion with the divine essence. It serves as a powerful tool for expressing love, surrender, and gratitude, leading one on the path of spiritual awakening and union with the ultimate reality.

Worship:

Worship in Bhakti Yoga without a personal God transcends external rituals and offerings directed towards a specific deity. Instead, it takes on a more abstract and symbolic form, focusing on cultivating an inner attitude of reverence and adoration towards the divine essence that pervades all creation.

Worship becomes a deeply personal and intimate expression of devotion, where one acknowledges the presence of the divine essence in every aspect of existence. Rather than worshiping a deity as an external entity, the practitioner recognizes the divine essence as immanent within themselves and all beings.

This shift in perspective transforms worship into an inner journey of self-discovery and spiritual realization.

The practitioner approaches every moment and every experience with a sense of reverence and awe, recognizing the sacredness inherent in all of creation.

In this case, worship is not limited to specific rituals or ceremonies but is woven into the fabric of everyday life. It encompasses acts of kindness, compassion, and selfless service (seva), performed with the awareness of serving the divine essence present in all beings.

It may also include contemplative practices such as meditation, where one reflects on the divine qualities of love, compassion, and wisdom that permeate the universe. Through meditation, the practitioner deepens their connection with the divine essence and cultivates a sense of unity and oneness with all of creation.

Symbolic representations such as sacred images, symbols, or natural elements may serve as focal points for worship, guiding one toward a deeper understanding of the divine essence. These symbols are not seen as mere representations of the divine but as manifestations of its presence in the world.

We find several examples of quotes relating to devotion in the Bhagavad Gita and below are a few of the most notable:

- "Those who worship Me with devotion, meditating on My transcendental form, to them I carry what they lack and preserve what they have." (Bhagavad Gita 9.22)

  This verse from the Bhagavad Gita underscores the importance of devotion and meditation on the divine form in Bhakti Yoga. Lord Krishna assures that those who worship Him with sincere devotion, focusing their minds on His divine form, will receive His grace. Krishna promises to fulfill the needs of devotees, providing what they lack and preserving what they already possess. This highlights the reciprocal relationship between the devotee and the divine, where worship and meditation lead to divine blessings and protection.

- "O son of Kunti, all that you do, all that you eat, all that you offer and give away, as well as all austerities that you may perform, should be done as an offering unto Me." (Bhagavad Gita 9.27)

  In this verse, Lord Krishna instructs Arjuna to perform all actions as an offering unto Him. This emphasizes the concept of worship in Bhakti Yoga as not limited to rituals or prayers but extending to every aspect of life. Krishna encourages Arjuna to cultivate an attitude of devotion and surrender by dedicating all actions, including mundane activities, offerings, and acts of charity, to the divine. Through such selfless offerings, the practitioner deepens their connection with the divine and advances on the path of Bhakti Yoga.

- "Worship of the Lord, which destroys all sins, is performed by one who has a disciplined mind, and has control over the senses, who has abandoned all desires and is free from attachment." (Bhagavad Gita 7.16)

> Here, we read about the qualities required for effective worship in Bhakti Yoga. Lord Krishna states that true worship, capable of eradicating sins, is performed by individuals with a disciplined mind and senses. The practitioner should have relinquished all desires and attachments, cultivating a state of inner purity and devotion. This highlights the importance of self-discipline, renunciation, and devotion as integral aspects of worship in Bhakti Yoga, leading to spiritual purification and divine grace.

Ultimately, worship in Bhakti Yoga without a personal God is an inner attitude of reverence, gratitude, and adoration towards the divine essence that transcends all forms and attributes. It is a continuous journey of self-discovery and spiritual awakening, where every moment becomes an opportunity to connect with the sacredness of life and realize the oneness of existence.

Surrender:

Surrender in Bhakti Yoga, irrespective of whether a personal deity is involved, is a fundamental aspect of the path of devotion. It entails relinquishing the ego, desires, and attachments to the divine essence that permeates all existence. This surrender is not an act of resignation or defeat but a conscious choice to align oneself with the divine will and trust in its wisdom and guidance.

At the heart of surrender is the recognition of the limitations of the egoic mind and the transient nature of worldly desires and attachments. It calls for an acknowledgement that the ego, with its sense of individuality and separateness, creates a barrier to experiencing the true essence of reality. By surrendering the ego, desires, and attachments, one opens themselves to the divine presence and allows it to guide their thoughts, words, and actions.

This surrender fosters a deep sense of humility, as one becomes aware of their dependence on the divine essence for guidance and support. It humbles the ego, recognizing that it is not the sole agent of action but merely an instrument through which the divine will operates.

Acceptance is another key aspect of surrender in Bhakti Yoga. The practitioner accepts the circumstances of life as manifestations of the divine will, whether they are perceived as favorable or unfavorable. This acceptance frees the mind from the burden of resistance and attachment, allowing it to remain open and receptive to the flow of life.

Surrender leads to a profound sense of inner peace and freedom. By letting go of the ego's need to control and manipulate outcomes, one can experience a deep sense of peace that transcends the fluctuations of the mind and emotions. This inner peace is not dependent on external circumstances but arises from the realization of one's innate connection with the divine essence.

Freedom is also inherent in surrender. By surrendering the ego, desires, and attachments, the practitioner liberates themselves from the bondage of ignorance (avidya) and the cycle of suffering (samsara). They recognize their true nature as eternal beings, beyond the limitations of the transient world, and experience a sense of liberation and expansiveness that transcends all boundaries.

Below are some quotes to give you an idea of what is required from a practitioner of Bhakti Yoga:

- "Those who surrender to the Self (Atman) are freed from the cycle of birth and death. Having realized the Self, they attain supreme peace." (Katha Upanishad 1.2.25)

  This quote emphasizes that surrendering to the realization of the Self (Atman) leads to freedom from the cycle of birth and death. By attaining self-realization, individuals achieve supreme peace. In essence, it highlights the transformative power of understanding one's true nature and transcending the cycle of reincarnation to attain inner tranquility.

- "To that universal Atman, consisting of consciousness, all beings offer themselves as objects of worship. Those who perceive themselves as separate from the Atman suffer in the cycle of birth and death." (Mundaka Upanishad 3.2.9)

  This verse speaks to the essence of reverence and surrender in Bhakti Yoga. It emphasizes the recognition of the universal Atman as the object of worship and the consequences of perceiving oneself as separate from it. Surrendering to the universal Atman leads to liberation from the cycle of birth and death.

- "Verily, the Self is beyond fear. One who has known this truth transcends sorrow. By the knowledge of the Self, the wise become free from desires, and attains bliss." (Chandogya Upanishad 7.1.3)

  This quote highlights the transformative power of self-knowledge in transcending fear and sorrow. It highlights that the realization of the Self leads to liberation from desires and attainment of bliss. In essence, it underscores the importance of understanding one's true nature to achieve inner peace and freedom from worldly attachments.

In summary, surrender in Bhakti Yoga is a profound practice that fosters humility, acceptance, inner peace, and freedom. It is a surrender of the ego to the divine essence, trusting in its wisdom and guidance, and aligning oneself with the flow of life. Through surrender, the practitioner experiences a deepening of their connection with the divine and realizes the true essence of their being.

Two further notions within Bhakti Yoga are that or spiritual transformation and union with the ultimate reality.

Spiritual Transformation:

Firstly, spiritual transformation unfolds as a mesmerizing journey of inner metamorphosis. Like a lotus emerging from the murky depths of a pond, practitioners undergo a remarkable process of purification and refinement. Through their unwavering devotion and heartfelt surrender to the divine, they shed the layers of ego and selfishness, allowing the radiant qualities of love, compassion, and selflessness to blossom within. This transformation is akin to a radiant sunrise illuminating the soul, bringing forth the divine virtues that lie dormant within every seeker.

- "When all desires that dwell in the heart are renounced, the mortal becomes immortal, and attains Brahman even here." (Chandogya Upanishad 8.1.6)

  This verse from the Chandogya Upanishad speaks to the transformative power of renunciation. By letting go of worldly desires and attachments, the practitioner realizes their true nature as immortal beings and attains union with Brahman, the ultimate reality.

Union with Ultimate Reality:

Secondly, the concept of union with the ultimate reality paints a captivating portrait of cosmic communion. It beckons practitioners to embark on a mystical voyage towards the boundless expanse of divine consciousness. Devotees aspire to dissolve the boundaries of individuality and merge seamlessly with the infinite ocean of divine presence. This union is not merely a destination but a transcendental odyssey, where the soul's yearning for oneness with the divine finds its ultimate fulfillment, and the symphony of creation resonates in perfect harmony.

- "That one, though formless, is called the form of forms; that one, though unthinkable, is called the thinker behind the mind; that one, though unperceivable, is called the perceiver of the

> mind; that one, though without support, is called the support of all." (Mundaka Upanishad 2.2.4)
>
> This verse from the Mundaka Upanishad elucidates the nature of the ultimate reality, which transcends all forms and attributes. It speaks to the ineffable nature of the divine essence, which can only be realized through profound spiritual practice and realization.

The path of Bhakti Yoga unfolds with boundless possibilities, even for those who don't resonate with a personal deity. Here, seekers immerse themselves in devotion to an abstract and universal divine essence, transcending the limitations of form and embracing the essence that permeates all of creation.

With hearts ablaze with love and reverence, practitioners embark on a mystical journey of surrender, relinquishing the ego's grasp and merging with the cosmic dance of existence. Through the sacred alchemy of devotion, they undergo profound spiritual transformation, shedding the shackles of individuality to become one with the pulsating rhythm of the universe.

In this divine symphony, Bhakti Yoga without a personal God becomes a gateway to union with the ultimate reality, where the boundaries of perception dissolve, and the soul merges effortlessly with the infinite. It's a journey that transcends words and concepts, leading seekers to the very essence of existence itself.

To sum up, Bhakti Yoga without a personal God beckons seekers to embrace the transformative power of devotion, offering a sacred path to spiritual realization and union with the ineffable beauty of the ultimate reality that transcends all forms and attributes.

Pray On

I was on a trip to Benares (an especially important holy city in Hinduism). This was pre-Instagram times so most people around me didn't know where I was visiting and what I was doing there. One evening during my stay in an Ashram, I met a couple from Germany who had just spent a year living in Rishikesh and were now going to Benares to spend some time there before heading back to Germany.

As we got closer and bonded over our shared interest in the Vedas, I learnt that the wife was doing a PostDoc in Physics at the Freie University of Berlin before she met her husband and discovered the world of Eastern spirituality. The husband was a musician, who discovered Eastern spirituality during his time spent in the company of some ISKCON devotees. One thing led to another and the couple convinced me to offer them some amateur classes in basic Sanskrit. The class that started as private tutoring for the couple quickly grew to include over a dozen people. One of the features of the lessons included some basic chants we would do together. Soon enough, it was my time to leave so we exchanged numbers and agreed to keep in touch.

I guess life is busy for everyone and we never really made contact after that.

Fast forward to very recently, I was speaking at a conference in Barcelona where I had multiple speaking engagements, including a book release and a keynote on the synthesis of science and spirituality. When I was on stage delivering a talk for my book launch, I noticed a pair of eyes in the audience looking at me keenly. The face seemed familiar although I couldn't exactly place how I knew the woman staring at me. The event was very rushed and as soon as it was over, I looked for her in the audience but couldn't find her, learning later she had dashed off to another session.

Later in the day, I delivered a keynote talk where I presented similarities in quantum physics, especially the string theory framework, and some of the esoteric teachings of the Vedas. The talk was quite eclectic, with enough science and math for the scientifically inclined and sufficient references to the Vedas to engage the spiritually inclined. My intention was to bridge the two sides. As I was leaving the stage, I noticed the same woman again in the audience but, this time, she waved to me. I waved back and suggested we meet in the foyer.

Once there, she quickly approached me and said, "Do you recognize me?" I apologized, saying I didn't. Then she said, "I am Zoya. Remember? Me and Mikin met you in Benares?" The penny immediately dropped! It was genuinely great to meet her again and we quickly exchanged new numbers, agreeing to meet for coffee soon. She had to rush off as she had a train to catch that evening for Madrid but she had stayed to come to my talk and speak to me.

We met a few months later in Berlin. They were both living in Hamburg and came up to Berlin to visit with me. It was a truly wonderful reunion and we reminisced on old times in India. Zoya excitedly told me that she attended the conference due to her daughter Ellie, who lived in Barcelona and was volunteering for the event.

After a while, Zoya asked me if she could ask me a question about my talk. I replied. "*Yes. Of course!*"

She then went on to enquire why I was trying to bridge science and spirituality, when it was clear that Vedas' wisdom was supreme. I was taken aback with that question and began by saying how interesting it was to get this question from someone who had a PhD in theoretical physics. I knew she had been doing work as part of her Postdoc on some dark matter candidates. We had a good chat about the subject but what stood out to me was her closing remarks. She said (I am reproducing my best recollection):

"Well, Leo. I was a little disappointed that in all your talks, you did not chant any shlokas or offer to have a Satsang at the event. Your talk was so scientific! But I guess it's OK. If, as a trained physicist, I can find peace in Satsang and prayer, and wear it as my life, it's only fair that as a Vedic scholar you adopt the paradigm I left behind to make sense of it with consciousness. Water must flow both ways."

I don't quite agree with her framing since, for starters, I am not a Vedic scholar by any stretch of the imagination but merely a student, and I was not as new to physics. In fact, I had been a keen student of all things quantum when I first met them, but we never discussed it due to meeting in a different context. Despite this, as I have thought about it more, I realize that there is some truth in what she said.

We must have some Bhakti Yogis who attempt to bridge Jnana Yoga, and vice versa, and this applies equally to Karam Yogis and other paths also! Only then is a true synthesis and richness of understanding and assimilation possible. And to that extent, I feel that, hopefully, our work with CMI is making some of that possible.

"Karma Yoga is about doing the right thing, not because of what you might get out of it, but because it is the right thing to do. It is the practice of living in harmony with the universe and fulfilling your purpose."

– Joseph Campbell

CHAPTER 9

KARMA YOGA: PATH OF SELFLESS ACTION

In the vast expanse of Vedanta philosophy, Karma Yoga shines as a beacon of selfless action and unwavering service. Rooted in the principle of detachment, it calls upon practitioners to perform their duties without expectation, surrendering the fruits of their labor to the divine orchestrator of the cosmos.

Karma Yoga is like a melody, where every action becomes an offering, a hymn of devotion sung to the divine essence that weaves through the tapestry of existence. With hearts filled with humility and minds steeped in surrender, practitioners navigate the labyrinth of life, recognizing each task as a sacred duty bestowed upon them by the divine.

At the heart of Karma Yoga lies the notion of selflessness, where the ego dissolves like mist in the morning sun, paving the way for the emergence of equanimity and inner peace. By relinquishing attachment to success and failure, pleasure and pain, practitioners traverse the path of purification, purging the mind of desires that obscure the radiant light of the soul.

But Karma Yoga is not merely a solitary pursuit; it is a call to embrace the interconnectedness of all beings and the pulsating heartbeat of the universe. Through selfless service, practitioners become agents of change, sowing seeds of compassion and kindness into the fields of society. With each act of service, they cultivate a deep sense of empathy and unity, transcending the illusory boundaries of ego and separateness.

As a spiritual discipline, Karma Yoga finds its roots deeply embedded in the philosophical underpinnings of Vedanta. The Bhagavad Gita, often considered the quintessence of Vedanta philosophy, serves as a cornerstone text for understanding Karma Yoga. Set against the backdrop of the battlefield of Kurukshetra, the Gita is a profound dialogue between Lord Krishna and the warrior prince Arjuna. Here, Lord Krishna expounds on the principles of Karma Yoga, outlining the path of selfless action, devotion, and spiritual wisdom.

In addition to the Bhagavad Gita, the Upanishads also contain profound insights into the principles of Karma Yoga. While they primarily focus on the metaphysical aspects of reality and the nature of the self (Atman) and Brahman, they also underscore the importance of righteous action performed with a spirit of detachment and devotion.

Furthermore, the Vedanta Sutras, also known as the Brahma Sutras, provide a systematic exposition of Vedanta philosophy, including the principles of Karma Yoga. Although they are more concerned with metaphysical inquiry and philosophical analysis, the Vedanta Sutras lay the groundwork for understanding the practical application of Karma Yoga within the context of spiritual evolution and liberation.

Overall, the sources for Karma Yoga in Vedanta are deeply rooted in the ancient scriptures of India, which serve as guiding beacons for seekers on the path of selfless action, devotion, and spiritual wisdom. They offer profound insights into the timeless principles of Karma Yoga and its relevance in the journey towards self-realization and liberation.

Below, we will walk through some of the main concepts of this life-changing approach.

Dedicating Actions to the Divine:

The practice of dedicating actions to the divine is a central tenet of Karma Yoga, serving as a powerful tool for spiritual transformation and union with the ultimate reality. It originates from the timeless wisdom of ancient scriptures and holds great importance in guiding practitioners towards selflessness, devotion, and inner peace.

We can discover profound spiritual insights into this practice in the Bhagavad Gita and the Upanishads. In each case, the practitioner of Karma Yoga sees every action as an opportunity to connect with the divine and serve the greater good. By dedicating their actions to the divine essence that pervades all of creation, they infuse their lives with purpose, meaning, and spiritual significance.

In the Bhagavad Gita specifically, Lord Krishna instructs Arjuna on the path of Karma Yoga, emphasizing the importance of performing actions as an offering to the divine:

- "Whatever a man sacrifices in this world, O Arjuna, is all offered to Me alone; of this there is no doubt." (Bhagavad Gita 9.27)

  This verse from the Bhagavad Gita stresses the importance of dedicating all actions to the divine. In Karma Yoga, the practitioner offers every action as a form of worship or service to the divine essence, recognizing that all actions ultimately belong to the divine.

- Examples:
  - Cooking a Meal:

- A practitioner of Karma Yoga may approach the task of cooking a meal with the intention of offering it to the divine. They may view the act of preparing food as a sacred ritual, infusing each step with love, care, and mindfulness. As they chop vegetables, stir pots, and season dishes, they offer every action as a form of worship, recognizing the divine presence in the nourishment that sustains life.

◊ Working in an Office:

- In the workplace, a practitioner of Karma Yoga may dedicate their professional duties to the divine. Whether it's attending meetings, responding to emails, or collaborating with colleagues, they approach each task with a sense of reverence and service. By viewing their work as an opportunity to contribute to the greater good and serve the divine essence present in all beings, they infuse their actions with purpose and meaning.

◊ Volunteering at a Homeless Shelter:

- When volunteering at a homeless shelter, a practitioner of Karma Yoga may see their service as an offering to the divine. As they serve meals, provide clothing, and offer companionship to those in need, they recognize the divine presence in each individual they encounter. Their actions are motivated by a desire to alleviate suffering and serve the divine essence that exists within every human being.

◊ Caring for a Loved One:

- In caregiving roles, such as caring for an elderly parent or a sick loved one, a practitioner of Karma Yoga may approach their responsibilities with devotion and surrender. They see the act of caregiving as a sacred duty entrusted to them by the divine. As they attend to the needs of their loved one with compassion, patience, and selflessness, they offer every action as a form of worship, recognizing the divine presence in the bond of love and compassion.

Performing Actions without Attachment:

Performing actions without attachment is not merely a mechanical execution of tasks but a divine offering — a sacred ritual that sanctifies even the most ordinary of actions with the fragrance of devotion and surrender. It is a profound practice of aligning one's will with the divine will, and relinquishing the ego's grip on the fruits of labor.

In Karma Yoga, every action becomes a brushstroke on the canvas of existence, painting the masterpiece of one's life with the colors of selflessness and detachment. Whether cleaning the roads or leading nations, planting seeds or running a business, each action is imbued with the sacred intention of service and offering, transcending the boundaries of ego and ushering seekers into the realm of divine grace.

Through the practice of performing actions without attachment, seekers come to realize that true freedom lies not in the pursuit of personal gain or recognition but in the surrender to the divine flow of life. Like a river flowing effortlessly toward the ocean, they surrender their desires and expectations, allowing the currents of divine grace to carry them toward the shores of enlightenment.

Performing actions without attachment is about engaging with the world with a sense of detachment and equanimity. It is about embracing life's challenges with courage and resilience, knowing that success and failure, pleasure and pain, are but fleeting moments in the grand tapestry of existence.

In the practice of performing actions without attachment, seekers come to realize the impermanence of the material world and the transient nature of worldly attachments. They recognize that true fulfillment lies not in the pursuit of external pleasures but in the realization of the eternal truth that dwells within.

We can find some examples of how this can be applied below:

- "One who sees inaction in action, and action in inaction, is intelligent among men, and he is in the transcendental position, although engaged in all sorts of activities." (Bhagavad Gita 4.18)

This verse highlights the essence of performing actions without attachment to the outcomes. In Karma Yoga, the practitioner performs actions with a spirit of detachment, recognizing that the results are ultimately beyond their control and surrendering them to the divine will.

- ♦ Examples:
  - ◊ Planting a Garden:
    - A practitioner of Karma Yoga decides to plant a garden in their backyard. They carefully select seeds, prepare the soil, and plant them with love and care. Despite their efforts, they recognize that the growth of the plants and the abundance of the harvest are ultimately beyond their control. They surrender the outcome to the divine will, trusting that whatever results arise will be in accordance with the greater plan.
  - ◊ Giving a Presentation:
    - In a professional setting, a practitioner of Karma Yoga is tasked with giving a presentation to their colleagues. They prepare diligently, research thoroughly, and practice their delivery. However, as they stand before the audience, they release any attachment to how the presentation will be received. They speak with sincerity and authenticity, trusting that the outcome is in the hands of the divine and surrendering

any expectations or desires for praise or recognition.

- ◊ Supporting a Friend:
    - A practitioner of Karma Yoga offers support to a friend who is going through a difficult time. They listen attentively, offer words of comfort, and provide practical assistance. However, they refrain from becoming emotionally invested in how their friend will respond or whether their situation improves. Instead, they offer their support unconditionally, recognizing that the outcome is beyond their control and surrendering it to the divine will.

In essence, the practice of performing actions without attachment in Karma Yoga is a sacred dance — a cosmic ballet that leads seekers from the shadows of ego to the radiant light of enlightenment. It is a journey of self-discovery and self-realization that promises not only personal liberation but the realization of the eternal truth that lies at the heart of existence.

Cultivating Selflessness and Detachment:

Cultivating selflessness in Karma Yoga is not merely an act of altruism but a profound recognition of the interconnectedness of all beings. It is a sacred vow to serve the world with an open heart and a compassionate spirit, knowing that every action, no matter how small, has the power to ripple through the fabric of existence and touch the lives of countless souls.

Through the practice of cultivating selflessness, seekers come to realize that true fulfillment lies not in the pursuit of personal gain or recognition but in the act of giving itself. Like a river flowing effortlessly towards the ocean, they offer their talents, their time, and their resources to uplift and inspire others, knowing that in the act of giving, they receive the greatest gift of all — the gift of divine grace.

Here again, cultivating selflessness is not about renouncing the world or withdrawing from life's responsibilities; it is about engaging with the world with a sense of detachment and equanimity. It is about performing one's duties and responsibilities with dedication and sincerity.

By practicing selflessness and detachment, one comes to understand the impermanence of the material world and the fleeting nature of worldly attachments. True freedom lies not in the pursuit of external pleasures but in the realization of the eternal truth that dwells within. Such gestures of love and devotion foster a deep sense of connection and harmony with all of creation.

- "He who knows the bliss of Brahman, from where words, together with the mind, turn away, not reaching it — knowing this bliss in the mind, let him not desire offspring, let him not desire wealth, let him not desire worlds; for him who desires the Self alone, all desires vanish." (Taittiriya Upanishad 2.7.1)

  This verse from the Taittiriya Upanishad speaks to the essence of cultivating selflessness and detachment. It emphasizes the transcendental nature of the bliss of Brahman and encourages the practitioner to renounce worldly desires in favor of realizing the true Self.

- Examples:
  - ◊ Volunteering at a Soup Kitchen:
    - A practitioner who seeks to cultivate selflessness and detachment volunteers regularly at a local soup kitchen. They devote their time and energy to serving meals to those in need, without expecting anything in return. By renouncing the desire for personal gain or recognition, they embody the spirit of selflessness and detachment, focusing solely on the well-being of others and the service of the divine essence present in all beings.
  - ◊ Donating to Charity:
    - Another example could be a person who practices detachment by regularly donating to charitable organizations or causes. They contribute financially to support initiatives that alleviate suffering, promote education, or protect the environment, without attachment to how their donations will be used or whether they will receive any recognition. Their actions stem from a genuine desire to serve the greater good and make a positive impact in the world,

rather than seeking personal gratification or reward.

◊ Engaging in Seva (Selfless Service):

- Seva is selfless service performed without any expectation of reward or recognition. This could involve activities such as cleaning sacred spaces, preparing meals for the community, or offering assistance to those in need. By engaging in seva with a spirit of detachment, practitioners cultivate qualities such as humility, compassion, and selflessness, transcending the ego's desires and realizing the true Self beyond worldly attachments.

◊ Practicing Generosity in Daily Life:

- Practicing generosity in everyday interactions is another way to cultivate selflessness and detachment. For example, a person may offer a listening ear to a friend in distress, lend a helping hand to a stranger in need, or express gratitude and appreciation towards others without expecting anything in return. By renouncing the need for validation or reciprocation, they embody the teachings of the Taittiriya Upanishad and realize the inherent bliss of the true Self beyond worldly desires.

In essence, the practice of cultivating selflessness and detachment in Karma Yoga is a sacred pilgrimage — a cosmic journey that leads seekers from the shadows of ego to the radiant light of enlightenment. It is a journey of self-discovery and self-realization that promises not only personal liberation but the realization of the eternal truth that lies at the heart of existence.

Purification of the Mind:

Purification of the mind in Karma Yoga is not merely a cleansing of the surface but a profound excavation of the depths — a sacred excavation that unearths the hidden treasures of the soul and purifies the inner landscape from the stains of ignorance and illusion. It is a journey of self-discovery and self-transformation that leads seekers from the shadows of darkness to the radiant shores of enlightenment.

Through the practice of purifying the mind, seekers come to recognize the intricate tapestry of thoughts, emotions, and impressions that weave through the fabric of consciousness. Like a skilled gardener tending to the soil, they cultivate the garden of their mind with mindfulness and awareness, uprooting the weeds of negativity and nurturing the seeds of virtue and wisdom.

It is a process that brings greater compassion and understanding, transforming practitioners into vehicles for spiritual growth and self-realization. In this state of inner balance and equanimity, the waves of the mind no longer disturb the tranquil surface of the soul.

It is also a gesture of devotion and surrender to the divine will, with the recognition that the mind is a sacred vessel through which the divine expresses itself in the world. In fact, it is a sacred duty — a call to action that inspires seekers to radiate the light of purity and compassion into the world. In the luminous glow of purified minds, seekers become beacons of hope and inspiration, guiding others out of the darkness of ignorance and into the radiant embrace of truth and love.

- "The mind is indeed the source of bondage and liberation. By binding itself with attachment to objects, the mind brings bondage, while by ceasing such attachment, it attains liberation." (Amritabindu Upanishad 2)

  This verse from the Amritabindu Upanishad highlights the role of the mind in bondage and liberation. In Karma Yoga, the practitioner purifies the mind by gradually reducing attachment to worldly desires and cultivating a spirit of detachment and equanimity.

- Examples:
  - Meditation Practice:
    - A practitioner of Karma Yoga engages in regular meditation practice to purify the mind and reduce attachment to worldly desires. During meditation, they observe the fluctuations of the mind without getting caught up in them, cultivating a sense of detachment and equanimity. Through consistent practice, they gradually dissolve the mental patterns that bind them to the cycle of desire and suffering, experiencing greater inner peace and clarity.

◊ Mindful Consumption:

- Another example could be a person who practices Karma Yoga by adopting mindful consumption habits. They become aware of their desires and impulses, questioning whether their actions align with their deeper values and spiritual aspirations. By exercising discernment and moderation in their consumption of material goods, entertainment, and sensory pleasures, they gradually reduce the hold of worldly desires on their mind, leading to greater freedom and contentment.

◊ Service to Others:

- Engaging in selfless service to others is another way practitioners purify the mind in Karma Yoga. By directing their energy toward the welfare of others, they shift their focus away from self-centered concerns and desires, cultivating a spirit of generosity, compassion, and detachment. Through acts of kindness, empathy, and service, they transcend the narrow confines of the ego and experience a deeper sense of interconnectedness with all beings.

◊ Study of Sacred Texts:

- Studying sacred texts and teachings that emphasize the impermanence of worldly desires and the importance of spiritual realization can also help purify the mind in Karma Yoga. As practitioners reflect on the wisdom contained in these texts, they gain insight into the nature of the mind and its role in bondage and liberation. They develop a deeper understanding of the transient nature of worldly pursuits and the lasting fulfillment that comes from aligning with the divine will.

The practice of purifying the mind in Karma Yoga is a sacred pilgrimage of self-discovery and self-transformation that leads one from the depths of darkness to the heights of enlightenment. It is a journey that promises not only personal liberation but the realization of the eternal truth that lies at the heart of existence.

In summary, Karma Yoga is a transformative path that encourages selfless action, service, and detachment from the fruits of one's actions. It is a practice of offering all actions to the divine and performing them with a spirit of humility, surrender, and devotion. Through Karma Yoga, practitioners purify their minds, cultivate selflessness, and ultimately attain spiritual growth and liberation. Within this sacred tapestry of Karma Yoga is revealed the timeless truths that lie at the heart of existence. It is a journey that promises not only personal liberation but the realization of the eternal truth that dwells within each and every one of us.

What needs to be done

On the way back, Tangi asked Jason why he worked so hard at helping the homeless after his regular work hours. Didn't he feel tired at the end of the long day?

Jason replied, "*I don't do it because I want to do it or want something from it. I do it because I truly believe it needs to be done. Someone must do it, so why not me.?*

"But why does it always have to be you? I miss not being able to do fun things with you. Don't you want that, too?" Tangi persisted.

"But it's not about what I want. It's about what wants me, OK?" Jason replied.

Tangi understood that she may not fully appreciate what Jason was saying but he seemed certain, so she let it be.

Jason was left wondering, "Will it ever be done and will I be able to get on with regular life?" He didn't want to lose Tangi.

He asked himself, "Where, then, is the boundary in selfless action? How much is too much and how will I know?"

"Raja Yoga is the royal path to self-realization. It involves practices like meditation, concentration, and contemplation, leading to the direct experience of our true nature beyond the limitations of the ego."

– Alan Watts

CHAPTER 10

INNER KINGDOM: RAJA YOGA'S JOURNEY TO SELF-REALIZATION

As our voyage of spiritual exploration unfolds, Raja Yoga stands tall as the "royal path" –a majestic journey into the depths of consciousness and self-discovery. Like a grand library filled with endless aisles of words of wisdom, Raja Yoga beckons seekers to leaf through the pages of self-transcendence and inner mastery.

Raja Yoga holds practices that stretch beyond the physical realm — practices such as asana (physical postures), pranayama (breath control), concentration, meditation, and the pinnacle of spiritual attainment, samadhi (state of deep absorption). These practices, meticulously composed, form the intricate bibliography of Raja Yoga, guide practitioners on a sacred pilgrimage into the luminous realms of the mind and spirit.

Not merely an exercise in contorting the body into complex postures or controlling the breath, Raja Yoga is about delving into the recesses of the mind and unraveling the mysteries of existence. Through the practice of Raja Yoga, one strives to still the turbulent waves of the mind, quieting the incessant chatter of thoughts and emotions, and attain a state of deep absorption where the boundaries between the self and the universe dissolve like melting ice.

While Vedanta may not explicitly prescribe the practices of Raja Yoga, it acknowledges their profound significance as aids to self-inquiry. In the journey of self-discovery, meditation, and concentration serve as invaluable tools, guiding one into inner exploration and revelation.

The main sources of Raja Yoga draw from a rich history of ancient scriptures, philosophical texts, and enlightened sages who have illuminated the path of inner exploration and self-realization. One of the key sources and figures include Patanjali's Yoga Sutras. Patanjali, often regarded as the father of Raja Yoga, compiled the Yoga Sutras — a seminal text that systematically outlines the principles and practices of classical yoga.

A towering figure in the modern revival of yoga and Vedanta is Swami Vivekananda, who played a pivotal role in popularizing Raja Yoga in the West. Through his teachings and lectures, Vivekananda emphasized the practical aspects of Raja Yoga, advocating for its universal applicability and transformative potential.

Sri Aurobindo, a visionary philosopher, yogi, and spiritual leader, delved deeply into the practice of Raja Yoga as a means of individual and collective evolution. Through his integral yoga, Aurobindo integrated the insights of traditional yoga with modern psychology and philosophy, offering a holistic approach to spiritual transformation.

These influential thinkers, among others, have played a significant role in shaping the understanding and practice of Raja Yoga, guiding seekers on a transformative journey of inner exploration and spiritual awakening. In essence, Raja Yoga is a sacred alchemy — a divine synthesis of ancient wisdom and modern understanding, weaving together the threads of tradition and innovation into an epic tome of spiritual evolution.

Raja Yoga Practises

At this point, I want to take you a little deeper into the practices of this enlightening path.

1. **Asana** is one of the foundational practices of Raja Yoga, which refers to physical postures designed to promote strength, flexibility, and balance in the body. The term is derived from the Sanskrit word "āsana," which means "seat" or "posture." While the Upanishads themselves do not delve deeply into physical postures, they emphasize the importance of cultivating a healthy and balanced body-mind complex as a foundation for spiritual practice.

   For example, the Taittiriya Upanishad (2.1) speaks of the necessity of physical and mental discipline as prerequisites for spiritual realization:

   "Dama, self-control, is the foundation. Dama is the control of the mind. Dama is the control of the senses. Dama is the control of the breath. Dama is the control of desire. Dama is the control of anger. Dama is the control of the tongue. Dama is the control of the stomach. Dama is the control of the organs of action. Dama is the control of the heart." (Taittiriya Upanishad 2.1)

   In the context of the Taittiriya Upanishad, "dama" refers to self-control or mastery over one's mind, senses, desires, and actions. It encompasses restraining the mind from wandering, disciplining the senses to avoid distractions, and regulating desires and impulses. Dama is considered a foundational quality for spiritual aspirants as it creates the necessary inner discipline to pursue higher states of consciousness and self-realization.

2. **Pranayama**, or breath control, is another integral aspect of Raja Yoga. Through conscious regulation of the breath, practitioners learn to calm the mind, balance the *energy channels (nadis),* and enhance vitality.

   "Prana is verily the Supreme Brahman. From prana indeed are all these creatures born. By prana, when born, they live. Into prana they enter at death." (Prashna Upanishad 3.1)

   This verse highlights the vital importance of prana as the life force that sustains all living beings. It underscores the idea that prana is not only responsible for the creation and maintenance of life but also for the transition of beings into the afterlife. While this particular verse does not explicitly mention pranayama, it lays the foundation for understanding the significance of pranayama in yoga and spiritual practice.

   By regulating the breath through practices such as *alternate nostril breathing (nadi shodhana)* or *rhythmic breathing (sama vritti),* practitioners learn to calm the mind, balance the energy channels *(nadis*), and enhance vitality, preparing the ground for deeper states of meditation and self-realization.

3. **Focused Meditation (dharana), awareness meditation (dhyana), and samadhi (union with the divine)** are central practices in Raja Yoga aimed at achieving profound states of inner absorption and realization.

The Katha Upanishad (6.10-11) speaks of the importance of focus and awareness meditation in transcending the limitations of the ego and realizing the true Self:

"When the five senses are stilled, together with the mind, and the intellect ceases to be active: that is called the highest state. They say yoga is this complete stillness in which one enters the unitive state, never to become separate again." (Katha Upanishad 6.10)

This verse underscores the significance of calming the senses and quieting the mind in the practice of meditation. It highlights the state of complete stillness and unity that is attained when the distractions of the external world are transcended, allowing the practitioner to merge with the divine essence. Through disciplined practice, practitioners learn to focus their attention single-pointedly, gradually quieting the fluctuations of the mind and turning inward towards the source of consciousness.

4. **Samadhi** is the highest state of consciousness, often described as a state of profound absorption or union with the divine. It represents the pinnacle of yogic attainment, where the individual self merges with the universal consciousness, and all distinctions between subject and object dissolve.

The Upanishads describe four states of consciousness: waking (jagrat), dreaming (svapna), deep sleep (sushupti), and the fourth state (turiya) beyond the three.

In the waking state, individuals experience the external world through their senses, interacting with the physical environment. During dreaming, the mind creates a subjective reality, often disconnected from external stimuli, where subconscious desires and thoughts manifest in the form of dreams. In deep sleep, the mind rests, and individuals enter a state of unconsciousness, devoid of dreams or awareness of the external world.

The fourth state, turiya, transcends the limitations of the other three states. It is described as a state of pure consciousness, beyond waking, dreaming, and deep sleep. In turiya, individuals experience a profound sense of unity and interconnectedness with the universe, realizing their essential nature as divine beings. The Mandukya Upanishad refers to it in this way:

“That which is without letters (unmanifested), imperceptible, incomprehensible, the cessation of the phenomenal world, the auspicious, the non-dual — that is the fourth state (Turiya). This is Atman, and this has to be realized.” (Mandukya Upanishad 7)

Through deep meditation and contemplation on the nature of consciousness, practitioners penetrate the layers of illusion (maya) and realize their essential unity with Brahman, the unchanging reality underlying all phenomena.

By turning inward and exploring the depths of their own consciousness, seekers come to understand that the external world is but a manifestation of the divine within, and that true liberation lies in realizing this inherent unity with Brahman.

The concept of turiya and the exploration of consciousness are central themes in various Upanishads, including the Mandukya Upanishad, which extensively discusses the nature of consciousness and the states of samadhi.

Practical examples of Raja Yoga in everyday life

How are the principles of Raja Yoga applied in daily life? They may include the following:

- **Morning Meditation Practice**: A practitioner sets aside time each morning for meditation, sitting in a quiet space and focusing their attention inward. They may practice techniques such as practices of mindfulness and meditation, mantra chanting, or breath awareness to still the mind and deepen their connection to the divine essence within.
- **Yoga Asana Practice:** Incorporating a regular yoga asana practice into one's routine can help cultivate physical health, mental clarity, and emotional balance. By moving mindfully through a series of postures, practitioners align their body, breath, and mind, preparing themselves for deeper states of meditation and inner contemplation.

- **Breath Awareness** Throughout the Day: Practitioners of Raja Yoga may cultivate awareness of the breath throughout the day, using it as an anchor to stay present and centered amidst the busyness of life. They may take moments to pause and consciously deepen their breath, allowing it to calm the mind and bring them back to the present moment.
- **Retreats and Intensive Practices:** Engaging in periodic retreats or intensive practices dedicated to meditation and inner contemplation can provide opportunities for profound spiritual growth and insight. By immersing themselves in a supportive environment focused on spiritual practice, practitioners deepen their understanding of themselves and the nature of reality.

In essence, Raja Yoga offers a comprehensive framework for spiritual transformation and self-realization, incorporating practices that quiet the mind, expand consciousness, and lead to direct experience of the ultimate reality. While Vedanta may not explicitly prescribe the practices of Raja Yoga, it acknowledges their importance as aids to self-inquiry and awakening to the true nature of the Self.

The value of experiential practices in deepening one's understanding and realization is an inherent part of Vedanta philosophy. In that respect, Raja Yoga and Vedanta philosophy complement each other in the journey of spiritual awakening. While Vedanta provides the philosophical framework for understanding the nature of reality, Raja Yoga offers practical techniques for realizing it through direct experience. Together, they form a synergistic approach to self-realization and ultimate liberation.

To sum up this section on the specialist paths, Vedanta recognizes and accommodates various routes to transcendence, each offering a unique approach to spiritual awareness and liberation. While Vedanta is closely associated with Jnana Yoga, it acknowledges the validity of Bhakti Yoga, Karma Yoga, and Raja Yoga as complementary paths suited to different aspirants' temperaments and stages of spiritual evolution. Ultimately, the goal of all these paths is to transcend the limitations of the ego and acknowledge one's essential unity with the ultimate reality, Brahman.

As a side note, there is a growing body of literature supporting the beneficial effects of meditation and mindfulness practices on mental and physical well-being. Numerous studies have shown that regular meditation can reduce stress, anxiety, and depression, improve attention and cognitive function, and enhance overall psychological resilience.

Neuroscientific research has also demonstrated that meditation can induce changes in brain structure and function, leading to increased gray matter density in areas associated with attention, emotional regulation, and self-awareness.

Moreover, studies on the physiological effects of meditation have revealed reductions in markers of inflammation, stress hormones, and blood pressure, as well as improvements in immune function and cardiovascular health. You will find a list of studies at the end of this book for your further reading.

These findings suggest that Raja Yoga practices, such as meditation and breath awareness, may have profound effects on both mind and body, aligning with the principles of Vedanta philosophy and supporting the pursuit of spiritual growth and self-realization.

Not that Vedanta has anything to prove. As an ancient philosophical and spiritual tradition rooted in the wisdom of the Upanishads, it predates modern science by millennia. Its teachings and insights into the nature of reality, consciousness, and the self have been passed down through oral tradition and written texts for thousands of years. Therefore, Vedanta doesn't seek validation or approval from the scientific community, as its principles are not based on empirical experimentation or the scientific method.

Instead, Vedanta offers a comprehensive framework for understanding the human experience and the universe as a whole, drawing upon introspection, contemplation, and direct spiritual experience. It addresses fundamental questions about existence, purpose, and the nature of reality through philosophical inquiry and inner exploration rather than through external observation or experimentation.

While modern science and Vedanta operate in different domains and employ different methodologies, there is a growing recognition of the complementary nature of their insights. Science provides valuable knowledge about the physical world and its mechanisms, while Vedanta offers profound insights into the nature of consciousness, self-awareness, and the interconnectedness of all life.

Ultimately, Vedanta invites individuals to explore and experience its teachings firsthand through personal practice, meditation, and self-inquiry. Its wisdom transcends the limitations of time and culture, offering timeless truths that resonate across generations and cultures, irrespective of scientific validation.

Who has everything?

Summary of a conversation at a Gurukul (ancient school) between a Rishi and a disciple:

What would one want who has everything?

Freedom.

What would tie him down?

Karma.

And what exactly is the freedom that you are talking about?

Let me elaborate. Freedom from the known. Freedom from ignorance. Freedom from cycle of rebirth. Freedom from the limitations of self.

So how can we have this freedom?

By realizing the true nature of self, maya and the Brahman.

What, then, is the path?

There are many paths. You choose your own or make a new one.

The disciple went on to become a great Rishi himself, whose work is still being understood.

"The Hindu religion is the only one of the world's great faiths dedicated to the idea that the Cosmos itself undergoes an immense, indeed an infinite, number of deaths and rebirths."

– Thomas Huxley

PART III:

LIBERATION AND THE ETERNAL CYCLE

As we have discussed in the previous sections of the book, Vedanta, one of the oldest philosophical and spiritual traditions of India, delves deep into profound themes of liberation, samsara, and the eternal cycle. It offers profound insights into the nature of existence, the purpose of life, and the ultimate goal of human endeavor.

At its core, Vedanta explores the quest for ***liberation (moksha)*** from the ***eternal cycle of birth and death (samsara)*** through observing **ethical and moral principles (dharma)**, as well as the emphasizing concept of ***karma***.

Moksha represents the ultimate freedom from worldly constraints, allowing individuals to realize their true nature as the eternal Self. Seekers pursue this liberation through self-inquiry, meditation, devotion, and service, aiming to break free from the cycle of birth and death and attain eternal peace and fulfillment.

Samsara, the eternal cycle of birth, death, and rebirth, underpins Vedanta's teachings. Driven by karma, it embodies impermanence and suffering, perpetuated by ignorance of one's true nature. Vedanta illuminates the causes of human suffering, guiding individuals towards self-realization and liberation.

Dharma, the ethical principles of righteous living, plays a vital role in Vedanta. By adhering to dharma, individuals cultivate virtues such as truthfulness and compassion, leading to inner purification and spiritual growth.

Karma signifies the causal connection between actions and their consequences in Vedanta. It shapes the trajectory of one's spiritual journey, emphasizing the importance of aligning actions with ethical principles to understand suffering and attain liberation.

The concepts of moksha, dharma, samsara, and karma are not unique to Vedanta. You will find them within several other belief systems, albeit with subtle differences of understanding. Moksha, found in Eastern and Western mystical traditions, represents the ultimate liberation from the cycle of birth and death, leading to spiritual freedom, bliss, and union with the divine.

Samsara, shared by various Eastern religions like Hinduism, Buddhism, and Jainism, depicts the cyclic nature of existence driven by karma. This eternal cycle involves repeated births and deaths, shaping the spiritual journey of sentient beings.

Dharma, a fundamental principle in Hinduism and other spiritual traditions, encompasses moral and ethical principles governing individual conduct and societal harmony, emphasizing alignment with cosmic order and moral integrity.

What sets Vedanta apart in the realms of liberation, dharma, and samsara is its comprehensive and systematic approach to understanding the nature of reality and the human condition. Its holistic and inclusive nature makes it a unique and comprehensive system of spiritual wisdom in the exploration of liberation, dharma, and samsara.

In the chapters that follow, we will delve deeper into these profound themes and discover the timeless wisdom that leads to ultimate freedom and realization.

"Birth is the dawn of the ego. Death is the dusk of the ego. The sun never rises or sets. It is always shining. It is the earth that revolves around itself and round the sun."

– Swami Sivananda

CHAPTER 11

SAMSARA: THE ETERNAL DANCE OF BIRTH AND REBIRTH

In the vast depths of Vedanta lies the captivating concept of samsara – the timeless cycle of birth, death, and rebirth that envelops all sentient beings within the material realm.

It beckons us to contemplate the essence of existence itself, unveiling a profound truth: that life in this transient world is made up of impermanence and change. Yet, within this ceaseless flux, lies a sublime opportunity for growth and transformation.

Samsara, intertwined with the intricate web of karma, unveils the cosmic law of cause and effect, wherein our every action echoes through eternity, shaping the course of our journey through countless lifetimes. It is a reminder of the profound interconnectedness of all beings and the boundless potential that lies within each moment to shape our destiny and transcend the limitations of the material realm.

The word "samsara" originates from Sanskrit and is derived from from the prefix "sam" meaning "together" or "with," and the root "sara" meaning "to flow" or "to wander." This reflects the cyclic nature of existence, where beings are bound to the continuous cycle of birth, death, and rebirth, perpetually wandering through various states of existence together.

It's worth mentioning here that, while this definition of samsara as "wandering together" or "flowing together," is a common interpretation, there isn't a single specific source that provides this exact definition. Instead, it is an understanding derived from the etymology of the word in Sanskrit and its usage in various Hindu and Buddhist texts.

According to Vedanta, the soul (Atman) reincarnates from one body to another, experiencing the fruits of its past deeds and accumulating new karma along the way. This process continues indefinitely until the soul attains liberation (moksha) from the cycle of samsara.

The spiritual roots of samsara in Vedanta can be traced back to the Upanishads, which expound upon the nature of existence and the eternal quest for liberation. The Chandogya Upanishad, for example, compares samsara to a wheel continuously turning upon itself, perpetuating the cycle of birth and death until one breaks free from its grip through self-realization.

In the Bhagavad Gita, there is a dialogue between Lord Krishna and the warrior prince Arjuna, where Krishna discusses various philosophical concepts including samsara, karma, and the paths to liberation. The Brahma Sutras, also known as Vedanta Sutras, systematically summarize the philosophical teachings of the Upanishads, including discussions on samsara and liberation.

Prominent thinkers and spiritual teachers have also provided extensive commentary on the notion of samsara, such as the philosopher Adi Shankaracharya (8th century CE), theologian Ramanuja (11th to 12th century CE), and Madhvacharya (13th century CE), who was the founder of the Dvaita Vedanta school. More recently, contemporary thinkers such as Swami Vivekananda, Swami Sivananda, and Swami Chinmayananda have also contributed to the ongoing exploration and interpretation of samsara within Vedanta philosophy. While their perspectives may vary, they continue to offer valuable insights into the nature of existence, karma, and liberation.

Samsara is often depicted as a realm of *illusion (maya),* where beings are ensnared by the fleeting pleasures and transient experiences of the material world, mistaking them for lasting happiness and fulfillment. Maya obscures the true nature of reality and leads individuals to mistakenly perceive the impermanent as permanent and the unreal as real.

Within samsara, beings become entangled in a perpetual cycle of birth, death, and rebirth, driven by desires and attachments rooted in ignorance (avidya). They seek fulfillment and happiness in external sources such as material possessions, sensory pleasures, and worldly achievements, believing them to be sources of lasting contentment.

However, these sources of happiness are inherently unstable and incapable of providing lasting fulfillment. Despite their efforts, individuals remain trapped in a cycle of craving, attachment, and suffering, endlessly pursuing transient pleasures that ultimately lead to disillusionment and dissatisfaction.

The teachings of Vedanta emphasize the need to transcend the illusions of samsara and seek liberation (moksha) from the cycle of birth and death. This liberation is attained through spiritual realization (jnana), wherein individuals come to understand the impermanent and illusory nature of worldly phenomena. By cultivating self-knowledge, detachment, and devotion to the ultimate reality (Brahman), seekers can break free from the bonds of samsara and experience true inner peace and fulfillment beyond the realm of illusion.

Despite its inherent challenges and suffering, samsara serves as a vehicle for spiritual evolution and growth. Through the experiences gained in successive lifetimes, souls have the opportunity to learn, evolve, and ultimately transcend the cycle of birth and death. Vedanta teaches that the ultimate aim of human life is to break free from the bondage of samsara and realize one's true nature as the immortal, unchanging Self *(Atman),* which is identical with the ultimate reality *(Brahman).*

Envision a cosmic dance of souls, entwined in the eternal cycle of birth and rebirth known as samsara. Guided by karma, fueled by desires, it's a journey through the material realm, marked by both suffering and opportunity. But for all its challenges, samsara holds a promise: a chance for spiritual growth and realization. Within life's twists and turns, amidst its trials and triumphs, lies the path to awakening and liberation.

Let's delve deeper into the theme of samsara in Vedanta, referencing sacred texts for clarity and depth.

Concept of Samsara in Vedanta:

Samsara, as understood in Vedanta, refers to the perpetual cycle of existence wherein individual souls (jivas) undergo repeated births and deaths. This concept is rooted in the understanding of the soul's journey through various states of existence, driven by karma (the law of cause and effect) and desires.

The Chandogya Upanishad provides insight into this concept, stating, *"As a caterpillar, having come to the end of one blade of grass, draws itself together and reaches out for the next, so the self, having come to the end of one life and dispelled all ignorance, gathers in its faculties and reaches out from the old body to a new."* (Chandogya Upanishad 6.10.3)

Nature of Existence in Samsara:

Samsara is characterized by impermanence, suffering, and the continual flux of life and death. The Bhagavad Gita elucidates on the transitory nature of worldly existence, stating, "From the highest planet in the material world down to the lowest, all are places of misery wherein repeated birth and death take place" (Bhagavad Gita 8.16). This highlights the pervasive influence of samsara across all realms of existence.

Furthermore, the Upanishads also expound on the nature of samsara. The Katha Upanishad, for instance, explores the concept of death and rebirth, illustrating how the soul transmigrates through various states of existence until it attains liberation. It reflects on the impermanence of the physical body and the eternal nature of the soul, which undergoes the cycle of birth and death until it breaks free from the bondage of samsara.

Additionally, the concept of dukkha, or suffering, as elucidated in Buddhist philosophy, resonates with the Vedantic understanding of samsara. Both traditions recognize the inherent unsatisfactoriness and impermanence of worldly existence and offer paths to transcend suffering and attain liberation.

Karma and Samsara:

Karma plays a pivotal role in perpetuating the cycle of samsara. The actions performed by individuals accumulate karma, which determines the conditions of their future births. The Mundaka Upanishad emphasizes the connection between karma and samsara, stating, "According to their deeds and according to their knowledge, the embodied soul approaches a womb that is appropriate for it." (Mundaka Upanishad 3.1.4)

The concept of karma finds its roots in various Hindu scriptures, including the Vedas, Upanishads, and the Bhagavad Gita. These texts expound on the intricate workings of karma and its implications for individual souls navigating the cycle of samsara. The Bhagavad Gita, for example, extensively discusses karma yoga, the path of selfless action, and elucidates on the principles of karma and its role in spiritual evolution.

Furthermore, the Law of Karma is also a fundamental concept in other Indian spiritual traditions, such as Buddhism and Jainism. These traditions offer their perspectives on karma and its implications for individuals seeking liberation from the cycle of birth and death.

Overall, karma is a foundational principle in Vedanta philosophy, intimately connected with the cycle of samsara. Through the law of karma, individuals sow the seeds of their future experiences, shaping their journey through the realms of existence until they attain liberation from the cycle of birth and death.

Illusion of Maya in Samsara:

Samsara is often depicted as a realm of illusion (maya), where beings are deluded by the transient nature of worldly phenomena. The Katha Upanishad elucidates on the deceptive nature of samsara, stating, "From the unreal, lead me to the real! From darkness, lead me to light! From death, lead me to immortality!" (Katha Upanishad 1.3.14)

This quote beautifully encapsulates the seeker's longing to transcend the illusory nature of samsara and attain the ultimate truth. It expresses the aspirant's prayer to move from the unreal to the real, from darkness to light, and from mortality to immortality. This verse underscores the fundamental teaching of Vedanta, which is to seek liberation from the cycle of birth and death and realize one's true immortal nature beyond the realm of maya.

We learned in Chapter 5 that maya refers to the principle of illusion or cosmic ignorance that veils the true nature of reality. It creates the deceptive appearance of diversity, multiplicity, and separation in the world, leading beings to mistake the transient world of appearances for ultimate reality.

This concept is expounded upon in various Upanishads and Vedantic texts, including the Katha Upanishad, Brihadaranyaka Upanishad, and Mandukya Upanishad. These scriptures emphasize the illusory nature of the phenomenal world and the need for seekers to transcend maya through spiritual realization.

The notion of maya serves to remind seekers of the impermanence and fleeting nature of worldly phenomena, prompting them to seek the eternal and unchanging reality beyond the realm of *maya*.

In Vedanta philosophy, the realization of maya's illusory nature is a crucial step on the path to liberation. By recognizing the deceptive nature of the phenomenal world and turning inward towards the eternal truth, seekers can transcend the cycle of samsara and attain spiritual liberation.

Purpose of Samsara:

Despite its inherent challenges, samsara serves as a crucible for spiritual evolution and growth. Through the experiences gained in successive lifetimes, souls have the opportunity to learn, evolve, and ultimately transcend the cycle of birth and death. The Brihadaranyaka Upanishad reflects on the transformative potential of samsara, stating, "The soul that knows Brahman becomes Brahman indeed." (Brihadaranyaka Upanishad 4.4.6)

But how do we harness the power of samsara in our spiritual path to enlightenment? In the Vedanta school of thought, that involves cultivating awareness, practicing discernment, and aligning our actions with spiritual principles.

Here are some ways Vedanta encourages us to acknowledge samsara and strive for liberation:

- **Mindfulness and Awareness:** Cultivate mindfulness in your daily activities. By staying present and attentive to your thoughts, emotions, and actions, you can become more conscious of the patterns and tendencies that contribute to the cycle of samsara. Mindfulness helps you recognize the transient nature of worldly pursuits and the deeper longing for spiritual fulfillment.

- **Detachment from Desires:** Practice detachment from worldly desires and attachments. While it's natural to have preferences and aspirations, excessive attachment to material possessions, relationships, or outcomes can perpetuate the cycle of samsara. By cultivating inner detachment and contentment, you free yourself from the relentless pursuit of fleeting pleasures and find greater peace within.
- **Self-Reflection and Self-Inquiry:** Engage in regular self-reflection and self-inquiry. Take time to contemplate the deeper aspects of your existence, such as the nature of the self, the purpose of life, and the impermanence of worldly phenomena. Through introspection, you can gain insight into the underlying causes of suffering and develop clarity about your spiritual journey.
- **Practicing Karma Yoga:** Embrace the path of Karma Yoga, the yoga of selfless action. Perform your duties and responsibilities with sincerity, integrity, and compassion, without attachment to personal gain or recognition. By offering your actions to the greater good and aligning them with spiritual values, you purify your karma and gradually transcend the cycle of samsara.

- **Seeking Knowledge and Wisdom:** Pursue spiritual knowledge and wisdom from authentic sources. Study sacred texts, scriptures, and teachings that offer insights into the nature of reality, the self, and the path to liberation. By cultivating a deeper understanding of spiritual principles, you equip yourself with the tools to navigate samsara and overcome its inherent challenges.
- **Cultivating Virtues and Values:** Cultivate virtues such as compassion, kindness, humility, and forgiveness in your interactions with others. Uphold ethical principles and values that promote harmony, justice, and integrity in society. By living in alignment with these virtues, you create positive karma and contribute to the collective evolution of consciousness.
- **Practicing Meditation and Inner Contemplation:** Dedicate time to meditation, inner contemplation, and spiritual practices that quiet the mind and connect you with your innermost essence. Through meditation, you can transcend the fluctuations of the mind, experience moments of inner stillness, and glimpse the eternal reality beyond the cycle of samsara.

In Vedanta, samsara emerges as the eternal melody of birth and rebirth, a cosmic symphony composed by the keys of karma and desire. It unfolds as a grand opera of existence, where souls traverse the realms of time and space, propelled by the ceaseless rhythm of cause and effect.

Within the labyrinth of samsara, beings wander through the ever-shifting sands of impermanence, seeking solace amidst the fleeting pleasures and pains of worldly existence. Yet, beneath the shimmering facade of illusion, lies the profound truth of the soul's journey — a journey fraught with challenges, trials, and the relentless pursuit of fulfillment.

But amidst the uncertainty of samsara, there is promise of transcendence and liberation. It is through samsara that souls are refined, purified, and awakened to the eternal truths of existence. Each cycle of birth and rebirth becomes a stepping stone towards enlightenment, a pilgrimage of the spirit towards the ultimate union with the divine.

Thus, samsara is not merely a cycle of bondage; it is also a path of redemption — a sacred journey towards the realization of one's true nature and the attainment of ultimate freedom. In the end, the soul transcends the illusions of samsara, and merges into the boundless ocean of divine consciousness.

Life beyond this, and Alice

Alice was talking to Jael one evening when the conversation led her firstly to be very upset and then elated later on. Both of them had known each other for many years, starting out as friends and growing to become much closer, sharing a deep bond. As Alice changed careers and decided to become a fashion entrepreneur, Jael took on the role of mentor for her as well, having been a successful luxury executive himself.

One evening, after they had returned home from dinner, they continued to enjoy a relaxing night over a bottle of wine. Jael was feeling particularly vulnerable and shared with Alice that he had been told by a soothsayer during his travels through Asia that he would have a short life. It may come to pass that he would not be around for very long. This made Alice cry instantly as she could not imagine what that would mean for her. While they were not engaged, married, or even officially a couple, they shared a love that only they understood and wanted this bond to last a lifetime, irrespective of how life changed for either of them.

Jael was taken aback by this and pedaled back, saying he had only been told that but that he wasn't planning on dying anytime soon. He smiled at Alice while tenderly holding her hand. As the night went on, their conversations became deeper until, eventually, Alice asked Jael, "Do you believe there is life after this?" The question had special significance, as she knew that Jael had had deep spiritual experiences in the past and was also mystical in his own way. He responded by saying that, for him, the afterlife was as real as this life. He recognized that both experiences are still Maya, and that freedom lies beyond life and Samsara.

"*Can we find each other in the afterlife?*" Alice asked.

"Yes, if we make a karmic pact, we can," replied Jael.

"OK. Then let's do this," Alice responded. *'If you leave early, you will find me when I arrive. If I leave early, I will find you when you arrive. Deal?'*

"But this assumes we only have life only on this planet. What if one of us is sent to another planet – let's say, Pleiades?"

"Tell me more about Pleiades."

And they talked.

"Pleiades sounds like fun. Let's plan to meet again there."

Both nodded in agreement.

Jael was up late that night wondering what a powerful feat of imagination the idea of an afterlife was; that it can offer both powerful incentives to live a 'good life' and also hope that there is more to come for those who don't have it all in this life. *What if religions had invented the afterlife as a concept just for this purpose? What if it was merely about removing the fear of dying, giving hope for eternal peace?*

Being too tired to think any more, Jael concluded he would never know for sure if there is an afterlife and what it looks like. But he was very happy for the moments he had shared that evening with Alice, where they reaffirmed their desire to be together. *Maybe there is no afterlife, but the thought of seeing her again there makes this life even better for me. Perhaps that is the whole point.*

With these last thoughts on his mind, he drifted off to sleep.

"Vedanta teaches that consciousness is singular, all happenings are played out in one universal consciousness and there is no multiplicity of selves."

– Swami Vivekananda

CHAPTER 12

DHARMA: TIMELESS WISDOM FOR EVERYDAY LIFE

In the captivating world of Vedanta, the essence of dharma is intertwined with the intricate principles of karma and the ultimate pursuit of moksha. Here, in a framework that transcends the confines of a personal deity, dharma emerges as a guiding light, illuminating the path to understanding the intricacies of existence and our place within the cosmic order. Join me as we place a magnifying glass over the profound significance of dharma, revealing its timeless wisdom and profound implications on our lives and beyond.

Colloquially, in India especially, dharma is sometimes erroneously translated in the sense of religion. Having said that, the original reference to dharma in the Vedic context points to the cosmic order or moral law that governs the universe, guiding individuals in leading righteous and virtuous lives. I realize I am paraphrasing a lot here, as I don't wish to write 10,000 words on the subject. ☺

What I can say here is that dharma, by corollary and implication, encompasses the duties, responsibilities, and ethical principles that individuals are expected to uphold in accordance with their station in life, social roles, and moral obligations.

In addition, the notion of dharma is also closely linked to the concept of karma, which posits that every action performed by an individual has consequences, either in this life or in future existences. The laws of karma dictate that one's actions, intentions, and thoughts determine the course of one's life and future rebirths. Thus, adhering to dharma involves acting in alignment with moral and ethical principles to accumulate positive karma and avoid negative consequences.

Dharma in the Vedanta tradition is seen as an integral aspect of the individual's journey towards self-realization and liberation (moksha). Brahman, the supreme consciousness, is the underlying essence of all existence, including the individual self (atman). However, due to ignorance (avidya), individuals mistakenly identify themselves with their limited ego and fail to realize their essential unity with Brahman.

This is where dharma comes in, serving as a means to overcome ignorance and align one's actions with the inherent order of the universe. Apart from having moral duties, we are also obliged to pursue self-knowledge (jnana) and carry out spiritual practices (sadhana) aimed at realizing our true nature as Brahman. By adhering to dharma and fulfilling our responsibilities in the world, we can purify our minds, cultivate virtues, and gradually disentangle ourselves from the cycle of birth and death (samsara).

You may have heard dharma being mentioned in relation to Buddhism and wonder if it is the same thing as in Vedanta. While the two spiritualities share some similarities, there are also key differences. In Buddhism, dharma refers not only to moral principles but also encompasses the teachings of the Buddha, the natural law of existence (dhamma), and the path to enlightenment (nirvana). Unlike Vedanta, which acknowledges the existence of a transcendental reality (Brahman), Buddhism emphasizes the impermanent and interconnected nature of all phenomena, rejecting the notion of a permanent self (anatman).

The reference to the "wheel of dharma" (dharmachakra) in Buddhism symbolizes the Buddha's teachings being set in motion, representing the path to liberation from suffering (dukkha) and the cycle of birth and death (samsara). As the wheel turns, it signifies the continuous cycle of existence and the perpetuation of karmic consequences.

Overall, while both Vedanta and Buddhism recognize the importance of dharma in guiding ethical conduct and spiritual growth, they differ in their philosophical frameworks and interpretations of ultimate reality, highlighting the richness and diversity of Eastern philosophical thought.

What are the sources that talk about dharma?

In Vedanta spirituality, the rules of dharma are derived from a combination of scriptural teachings, philosophical principles, and practical wisdom passed down through generations of spiritual seekers and teachers. Some of the main sources include the Vedas, the Upanishads, the Bhagavad Gita, and the Brahmasutras. We also have many commentaries and interpretations by Vedanta scholars such as Adi Shankaracharya, Ramanuja, and Madhvacharya, who provide a comprehensive understanding of dharma within the broader framework of Vedanta philosophy.

Here are some key sources through which the rules of dharma are understood in Vedanta:

- **Upanishads**: The Upanishads contain profound insights into the nature of reality, the self, and the ultimate truth (Brahman). While they may not explicitly outline rules of conduct like the Smritis, they emphasize the importance of ethical living and righteous conduct as integral aspects of spiritual practice. For example, the Chandogya Upanishad teaches the importance of truthfulness, compassion, and self-control as essential virtues for spiritual seekers.

- **Bhagavad Gita:** The Bhagavad Gita, often referred to as the "essence of the Vedas," presents the teachings of Lord Krishna to the warrior Arjuna on the battlefield of Kurukshetra. In the Gita, Lord Krishna expounds on the concept of svadharma (one's own duty) and the principles of righteous action (karma yoga). Arjuna is encouraged to fulfill his duty as a warrior, upholding dharma while remaining detached from the fruits of his actions.
- **Smritis and Dharmashastra**: While Vedanta primarily focuses on philosophical inquiry and spiritual realization, it acknowledges the importance of societal and moral norms upheld in texts like the Manusmriti and other Dharmashastra texts. These scriptures provide guidelines on social duties, personal conduct, and ethical behavior, which are seen as conducive to spiritual growth and harmonious living.
- **Guru Parampara:** In the tradition of Vedanta, the guidance of a guru (spiritual teacher) is paramount. The guru imparts not only philosophical knowledge but also practical wisdom on how to lead a life of dharma. Through personal instruction, example, and guidance, the guru helps disciples understand and internalize the principles of dharma in their daily lives.

- **Personal Reflection and Self-inquiry:** Vedanta encourages individuals to engage in self-inquiry and introspection to discern the principles of dharma in their own lives. By reflecting on the teachings of the scriptures, contemplating the nature of the self, and observing the workings of the mind, seekers strive to align their thoughts, words, and actions with the principles of truth, righteousness, and spiritual upliftment.

While the Upanishads do mention the concept of dharma, it's not as extensively discussed as in other scriptures like the Bhagavad Gita or the Vedas. In the Upanishads, dharma is often referred to in the broader sense of cosmic order or righteousness rather than specific moral or ethical duties.

For example, in the Chandogya Upanishad, there's a passage that speaks about the importance of dharma in maintaining the order and balance of the universe:

"One should not neglect duty, wealth, pleasure, and liberation." (Chapter 6, Section 1, Verse 2, Chandogya Upanishad)

Similarly, in the Brihadaranyaka Upanishad, there are discussions about dharma in the context of cosmic principles and the moral order of the universe. These philosophical insights contribute to a broader understanding of dharma as a guiding principle in Vedanta.

The Brihadaranyaka Upanishad, for example, explores the notion of karma as the driving force behind the cycle of birth and death (samsara) and emphasizes the importance of ethical conduct and selfless action.

One notable quote is:

"May our learning be luminous and purposeful."

While this specific quote may not explicitly mention dharma, the Brihadaranyaka Upanishad contains numerous teachings and discussions about the nature of dharma, ethical conduct, and the pursuit of righteousness.

The Mundaka Upanishad also elucidates on the importance of ethical conduct, stating that individuals attain appropriate embodiments based on their deeds and knowledge:

"That which is, O Svetaketu, is the Sat (the true essence); that is the Atman; that thou art, O Svetaketu. Thereupon, the other (Asvin) spoke unto him, to him, Yâgñavalkya: 'We shall explain to thee, O good looking one, Dharma (virtue). Listen to it; and let me speak. If you practice it, you will live; and if you do not, you will not live.'" (Chandogya Upanishad, 6, 8, V7)

In a godless universe, where the emphasis is on the impersonal and non-theistic aspects of reality, dharma serves as a guiding principle for moral conduct and spiritual growth. It provides a framework for individuals to cultivate virtues such as compassion, honesty, integrity, and selflessness, irrespective of the presence or absence of a divine being.

The Bhagavad Gita also underscores the importance of performing one's duty *(svadharma)* without attachment to the results, as a means to uphold dharma and attain spiritual liberation. It reads:

"You have a right to perform your prescribed duties, but you are not entitled to the fruits of your actions. Never consider yourself to be the cause of the results of your activities, nor be attached to inaction." (Chapter 2, Verse 47)

Furthermore, dharma is intrinsically linked to the pursuit of moksha, which refers to liberation from the cycle of samsara and the realization of one's true nature as eternal and divine. While the attainment of moksha may not necessarily depend on the worship of a personal God, it requires individuals to live in harmony with dharma, cultivate self-awareness, and engage in practices such as self-inquiry, meditation, and spiritual disciplines.

The Katha Upanishad delves into the nature of moksha as the transcendence of worldly attachments and identification with the immortal Self, leading to everlasting peace and liberation. One example quote is:

"Two birds, inseparable friends, cling to the same tree. One eats the fruits of the tree, while the other looks on without eating." (Katha Upanishad 2.1.1)

The verse, with the metaphor of the two birds, is deeply related to the concept of moksha (liberation). It symbolizes the relationship between the individual self (Jivatman) and the supreme reality (Paramatman), and elucidates the path to liberation from the cycle of birth and death (samsara). The tree represents the material world or the physical body, which is the realm of worldly experiences. The individual soul (Jivatman) is depicted as one of the birds, actively engaged in consuming the fruits of worldly pleasures and experiences. These fruits represent the results of karma (actions) accumulated throughout various lifetimes.

What does dharma mean in practical terms?

Dharma is a guiding principle that encourages individuals to align their actions with moral integrity, harmony, and the greater good. It fosters a sense of interconnectedness, respect for all life forms, and the recognition of a higher purpose beyond individual desires and interests. In the East, dharma guides millions of individuals to fulfill their duties and responsibilities in various spheres of life, including family, society, and the wider world.

To convey the essence of dharma to a Western audience, one might evoke parallels with concepts akin to moral principles or the immutable laws of nature. Just as the cosmos abides by the laws of physics, dharma stands as the moral compass navigating the vast expanse of human behavior and interaction.

In reality, dharma plays a fundamental role in shaping everyday life, manifesting itself in various ways:

Personal Conduct: Dharma guides individuals in their personal conduct, emphasizing virtues such as honesty, compassion, integrity, and self-discipline. It encourages individuals to act ethically and morally in all aspects of life, fostering harmony and balance within oneself.

Social Relationships: Dharma governs social relationships, including those within families, communities, and society at large. It outlines duties and responsibilities towards parents, children, spouses, teachers, and other members of society, promoting respect, cooperation, and mutual support.

Occupational Duties: Dharma prescribes duties and responsibilities associated with one's occupation or profession. Whether as a teacher, farmer, merchant, warrior, or laborer, individuals are expected to perform their duties diligently and ethically, contributing to the welfare of society while upholding moral principles.

Justice and Governance: Dharma informs principles of justice, governance, and law enforcement. Rulers and leaders are expected to govern with fairness, integrity, and compassion, upholding dharma in their decisions and actions to ensure the well-being and prosperity of their subjects.

Environmental Stewardship: Dharma extends to the relationship between humans and the environment, emphasizing the importance of ecological balance, sustainability, and reverence for nature. It encourages responsible stewardship of the earth and its resources, promoting harmony with the natural world.

Spiritual Growth: Dharma serves as a pathway to spiritual growth and liberation (moksha). By following one's duties and responsibilities in accordance with dharma, individuals purify their minds, cultivate virtues, and gradually transcend egoic desires, ultimately leading to self-realization and union with the divine.

Overall, dharma permeates every aspect of life, providing a moral and ethical framework for individuals to navigate the complexities of existence with integrity, purpose, and righteousness. It fosters a sense of interconnectedness, harmony, and fulfillment, guiding individuals on the path toward spiritual evolution and inner peace.

Examples of dharma in action can be found below:

- A teacher upholds their dharma by imparting knowledge and wisdom to students, nurturing their growth and development.
- A parent fulfills their dharma by providing love, guidance, and support to their children, ensuring their well-being and upbringing in line with moral values.
- In the broader social context, individuals uphold dharma by adhering to principles of honesty, compassion, justice, and respect for others, thereby fostering harmonious coexistence and social cohesion.
- Moreover, dharma extends beyond human interactions to encompass our relationship with the environment and all living beings, urging us to protect nature's gifts.
- By practicing ahimsa (non-violence) and environmental stewardship, individuals honor their dharma by respecting and preserving the sanctity of life in all its forms.

Think of dharma as a guiding light that illuminates the path of righteous living and moral conduct, guiding you to lead a life of integrity, virtue, and service to humanity and the cosmos. Through the observance of dharma, you can align yourself with the cosmic order and contribute to the evolution of consciousness towards higher ideals of truth, righteousness, and spiritual realization. That's not a bad perspective to have!

Incorporating these profound teachings from the Upanishads and Vedas into daily life, individuals can strive to live in harmony with dharma, cultivate positive karma through righteous actions, and ultimately attain the sublime goal of moksha. By aligning their lives with these timeless principles, seekers on the path of Vedanta philosophy aspire to realize their true nature and attain liberation from the cycle of birth and death.

In Vedanta spirituality, the rules of dharma are not rigid dictates but guiding principles that evolve in the context of one's spiritual journey. Ultimately, the practice of dharma is seen as a means to purify the mind, cultivate virtues, and realize one's essential nature as the eternal Self (Atman) beyond the transient realm of worldly existence.

Dharma is woven with threads of honesty, compassion, integrity, and social responsibility — cornerstones of a life lived in harmony with the universe's grand design. These virtues, revered across cultures and epochs, form the bedrock of a righteous and fulfilling existence, guiding individuals on a journey of self-discovery and altruistic action.

With each step we take in alignment with our dharma, we become architects of positive change, crafting a brighter future for all. It is in the moments of selfless service, the acts of kindness and compassion, that we truly embody the essence of dharma. In nurturing the seeds of goodness within ourselves, we sow the seeds of transformation in the world around us.

You could say, dharma is not just a path to personal fulfillment but a sacred quest to uplift humanity and embrace the divinity within us all!

Mine, Yours and Ours

I was once on stage with an Atheist debating the *nature of consciousness*. I don't typically engage in debates of a spiritual nature, mostly because I believe in the tradition of *"shastrartha."* This can be translated as a shared inquiry with an element of making a deeper meaning through recursive dialogue anchored in ancient wisdom. I have often felt that debates sometimes devolve into arguments, a judgment I guess emanates from the very many debates I participated in while in school as part of the national award winning debating team.

However, I accepted this particular invitation as the person I was debating with was a true scholar whom I respected very much, and I felt I would benefit a lot from his perspective!

The event took place in the presence of an audience of hundreds of people, and was interspersed with clapping and laughing as both of us made our points.

Very soon, the "temperature" of the arguments rose, and we fell into a scholarly pattern of taking humorous potshots at each other, both to make a point and aalso to engage the audience. At one point, the atheist scholar called me, *"Someone who, when pressed for empirical basis, would quickly abandon the scientific method and jump to pseudoscience and faith."* This got some cheers from the crowd. I responded in my rebuttal by addressing the atheist as, *"Someone who attempts to discredit for discrediting's sake, all knowledge and experiences that lie beyond his personal limitations, just as religious dogmatists discredited scientists in the past."* This also received applause.

Soon, the debate came to an end and we were both thanked by the organizer. The room slowly began to empty, with just a few people milling around looking for *"selfies"*.

The atheist speaker asked me what my plans were for the evening and I told him I had no specific plans. He asked if I would like to join him for dinner at his home, to which I immediately agreed. He suggested we drive together in his car as I had taken an Uber to the venue, and that perhaps I could provide him some feedback on the ideas for a book he was planning. I enthusiastically said, "Of course!"

One of the organizers was watching us and quickly commented on how surprised she was that my co-speaker and I were intending to spend the evening together, and that he was seeking my advice for his book. On asking why she was so surprised, she said we were just so aggressively attacking each other in the debate that she didn't think we would even have anything in common to discuss. We were both quite amused at that and smiled back at her. I then asked my fellow debater if he wanted to respond or if I should. He suggested, smilingly, *"You are the philosopher, Professor. You should unburden the young lady."*

So I told her, *"We weren't attacking each other, we were simply challenging each other's ideas. We have mutual respect and learn more about our point of view when it's challenged by others. It's actually a healthy scholarly tradition. We just disagree, we are not personally disagreeable."*

The atheist added, *"Although I am not entirely sure he is wrong, I am not sure he is as right as he thinks he is."*

The organizer went on to ask why, then, we debated with such passion. I replied by saying, *"We are both teachers. It's our dharma to clarify, educate, demystify concepts but also be open to other ideas different from our own."*

Suddenly her face lit up and she said, *"I need to go home and talk to my parents. We have been arguing a lot about same sex marriages. I just realized we may both be right or neither of us may be right. But we could and should just listen to each other. It may just be that we might learn from each other!"*

We both smiled and shook hands with her as we parted.

Last week, I received a LinkedIn message from her, sharing pictures of herself and her wife at their wedding with her family. She wrote, *"To be open to ideas of others should be a shared dharma. Thank you."* I couldn't agree more!

"Moksha is the realization of our own divinity. It is not something to be attained; it is something to be realized."

– Swami Rama Tirtha

CHAPTER 13

MOKSHA UNVEILED: LIBERATION BEYOND THEISTIC BOUNDARIES

In Vedanta, moksha stands as the shimmering jewel, the culmination of the soul's journey through the cosmos. It is more than a mere concept; it is the cosmic symphony's crescendo, the ultimate finale of the human spirit.

Moksha whispers to us of liberation, of emancipation from the shackles of illusion that bind us to the cycle of birth and death, to the ceaseless dance of samsara. It beckons us to transcend the limitations of our mortal existence, to soar beyond the confines of time and space, and to bask in the radiance of divine consciousness.

But moksha is not a distant dream or a far-off destination; it is the pulsating heartbeat within us, the spark of divinity waiting to be ignited. It is the journey inward, the quest to peel away the layers of conditioning and ego, to uncover the jewel of the true self hidden beneath.

Moksha is the sacred union of the individual soul (Atman) with the cosmic reality (Brahman), the realization that in essence, we are not separate from the divine, but are one with it. It is the ecstatic dance of unity, the blissful merging of the finite with the infinite.

And so, moksha is not merely an intellectual understanding but a direct experiential realization, a state of being where the boundaries of self dissolve, and we become immersed in the boundless ocean of consciousness. It is the ultimate freedom, the liberation of the soul from the cycle of suffering, and the emergence into the infinite expanse of divine love and wisdom.

To gain a deeper understanding of what moksha means according to Vedanta, we will explore some of its key concepts below.

Freedom from Samsara: Imagine you are entangled in the eternal cycle of birth, death, and rebirth — this is samsara, the relentless rhythm of existence. But amidst this endless cycle, there exists a profound longing for liberation, for escape from the repetitive patterns of life.

Moksha emerges as the radiant beacon of hope, illuminating the path to ultimate freedom. It whispers promises of transcendence, of breaking free from the chains of samsara and soaring into the boundless realms of the spirit.

Through spiritual practice and self-realization, individuals embark on a journey of inner transformation, shedding the veils of ignorance that obscure their true nature. With each step on this sacred quest, they inch closer to the elusive shores of liberation. In that respect, moksha is not just an abstract concept but a visceral experience — a liberation of the soul from the confines of worldly existence. It is the ecstatic moment of realization when the individual recognizes their inherent divinity and merges with the cosmic consciousness.

The Chandogya Upanishad (6.14.2) states, *"When a person knows Brahman, he becomes free from the cycle of birth and death."*

In other words, when individuals attain knowledge of Brahman — the ultimate reality — they transcend the limitations of the material world and break free from the cycle of samsara, achieving eternal freedom.

Realization of the Self: Embarking on the journey of moksha is akin to uncovering a hidden treasure within oneself — the realization of the true Self. It's a revelation that transcends the mundane and delves into the profound essence of existence.

The seeker comes to understand that their very essence — the Atman — is not merely a fleeting identity tied to the material world but is, in fact, a divine spark intricately woven into the fabric of the cosmos. It's a cosmic awakening, a recognition that the true Self is not bound by the limitations of time and space but is eternal, unchanging, and intimately connected with the ultimate reality — the all-encompassing Brahman.

In this realization, the seeker finds liberation from the shackles of illusion, stepping into the radiant truth of their divine nature. It's a journey of self-discovery that leads to profound transformation — a transformation that liberates the soul from the chains of mortality and unveils the eternal essence that lies within.

The Katha Upanishad (2.1.10) declares, "*When the five senses and the mind are still, and reason itself rests in silence, then begins the path supreme. This calm steadiness of the senses is called Yoga. Then one should become watchful, because Yoga comes and goes.*"

The quote emphasizes the importance of inner stillness, mindfulness, and vigilance in the practice of Yoga. It highlights that true Yoga is not just about physical postures or exercises but involves the mastery of the mind and senses, leading to the realization of the divine within.

Freedom from Ignorance (Avidya): In the grand narrative of spiritual liberation, moksha emerges as the triumphant hero, wielding the sword of enlightenment to vanquish the shadows of ignorance. It is more than just a fleeting escape; it is the grand unveiling of truth amidst the labyrinth of illusion.

In this cosmic path to realization, the seeker comes to recognize their true nature as pure consciousness — an essence that transcends the boundaries of time, space, and individual identity. It's a homecoming, a return to the divine essence that has always resided within.

With the dispelling of ignorance comes the dissolution of barriers — the barriers that divide, that segregate, that perpetuate the illusion of duality. Moksha is the ultimate liberation, the freedom to soar beyond the limitations of perception and bask in the radiant truth of unity.

The Brihadaranyaka Upanishad (4.4.6) states, *"He who knows that supreme Brahman becomes Brahman indeed."*

The quote implies that one who realizes the supreme reality, Brahman, becomes Brahman itself. In the context of moksha, it suggests that true liberation is attained through the direct realization of Brahman, where the individual merges with the infinite expanse of cosmic consciousness.

Transcendence of Dualities: Moksha is the ultimate act of transcendence — a journey beyond the realms of duality that govern the material world. It's a profound exploration of the cosmic realm, where the seeker untangles the intricate knots of opposites to uncover the timeless truth that lies beyond.

Moksha beckons one to rise above the ever-shifting tides of pleasure and pain, to transcend the dichotomy of good and evil, and to pierce through the veil that separates birth from death. It's an odyssey of the soul, a quest to unravel the mysteries of existence and discover the unchanging, blissful reality that lies at the heart of creation.

It's a journey of self-discovery, where the illusions of the material world dissolve, and the seeker finds solace in the eternal sanctuary of the soul. It's a liberation that transcends the boundaries of perception, leading one to the sublime realization of an unchanging reality that transcends the flux of the phenomenal world.

The Bhagavad Gita (2.15) teaches, *"That which is non-existent can never come into being, and that which is can never cease to be."*

The extract suggests that the true self (Atman) is eternal and unchanging. In the context of moksha, the essence of the self transcends the cycle of birth and death, leading to liberation (moksha) through the realization of its eternal nature.

How is moksha achieved?

How does one go about acquiring a direct experiential realization of one's true nature as divine and eternal in daily life? It sounds like a daunting task and one you may find impossible to achieve. Well, in Vedanta, experiencing moksha involves transcending the limitations of the ordinary mind and directly realizing one's true nature as divine and eternal. It requires a shift from intellectual understanding to direct experiential realization.

Here are some steps one can take to experience moksha:

- **Self-Inquiry:** Engaging in self-inquiry is like embarking on a practical adventure — a journey into the depths of our own being to uncover profound truths about existence. It's about rolling up our sleeves and asking ourselves some fundamental questions: Who am I? What is the nature of my consciousness? What is the source of my existence?

  This process isn't about lofty philosophical ponderings but about getting down to the nitty-gritty of our everyday experience. It's about taking moments throughout our day to pause and reflect, to observe our thoughts, feelings, and sensations with a curious mind.

  As we delve into these questions, we may find ourselves uncovering insights that challenge our assumptions and expand our understanding of ourselves and the world around us. We may come to see that our sense of identity is not as fixed as we once thought — that beneath the surface lies a deeper essence that connects us to something greater than ourselves.

  Practical self-inquiry involves experimenting with different techniques, including engaging in deep conversation with ourselves, to explore the nature of our consciousness and its relationship to the world. It's about carving out time in our busy lives to sit quietly and listen to the whispers of our inner voice, to tune into the subtle rhythms of our inner world.

Through this process, we begin to cultivate a deeper sense of self-awareness and understanding. We start to recognize the patterns and habits that shape our thoughts and behaviors, allowing us to make more conscious choices in our lives.

Ultimately, practical self-inquiry enables us to live more authentically and intentionally and is a practical tool for personal growth and transformation. It's this journey that can lead us closer to realizing our true nature and experiencing a deeper sense of fulfillment in our lives.

- **Meditation Practice:** Dedicating time to meditation practice is like carving out a quiet sanctuary amidst the hustle and bustle of everyday life — a space where we can retreat from the noise of the world and turn our attention inward. It's about setting aside a few moments each day to sit in silence, to quiet the chatter of the mind, and to reconnect with the source of our consciousness.

  In the practice of meditation, we learn to observe the fluctuations of the mind without getting caught up in them. We become like silent witnesses, watching as thoughts come and go like clouds passing over the sky. With each breath, we anchor ourselves in the present moment, allowing ourselves to sink deeper into a state of stillness and presence.

  As we cultivate this inner quietude, we may begin to experience glimpses of something deeper — an eternal aspect of our being that transcends the realm of thoughts and emotions. It's like catching a fleeting glimpse of the vast horizon stretching out above the surface of a turbulent sea — a reminder of the boundless expanse of our true nature.

In these moments of stillness, we may find ourselves tapping into a sense of peace and contentment that lies beyond the ups and downs of everyday life. We may feel a sense of connection to something greater than ourselves — a source of wisdom and guidance that resides deep within.

Meditation practice, then, is not just about achieving a state of blissful tranquility, but about cultivating a deeper relationship with ourselves and with the world around us. It's about learning to navigate the ebb and flow of life with grace and equanimity, rooted in the timeless wisdom of our own innermost being.

- **Contemplation of Sacred Texts**: Immersing ourselves in the study and contemplation of sacred texts is like embarking on a journey through the treasure troves of ancient wisdom — a journey that can profoundly enrich our understanding of spiritual principles and insights. It's about diving deep into the timeless wisdom passed down by enlightened sages and allowing their teachings to illuminate our path.

  As we delve into sacred texts such as the Upanishads, we open ourselves to a wealth of spiritual knowledge and guidance. We explore profound truths about the nature of existence, the essence of consciousness, and the ultimate reality that underlies all of creation. Through careful study and contemplation, we begin to unravel the mysteries of the universe and gain deeper insights into the workings of our own minds and hearts.

Each verse and teaching becomes a beacon of light, guiding us on our journey of self-discovery and spiritual growth. We may find ourselves drawn into a deeper understanding of our own innermost being, as we uncover hidden truths that resonate with the very core of our being.

Ultimately, the study and contemplation of sacred texts is not just an intellectual exercise, but a transformative journey of the soul. It's about opening our hearts and minds to the timeless wisdom of the ages and allowing it to permeate every aspect of our being. Through this process, we may find ourselves awakening to a deeper sense of purpose and meaning, as we journey ever closer to the truth that lies at the heart of existence.

- **Cultivation of Selflessness:** Cultivating selflessness is like tending to the garden of the soul — a practice that nourishes our inner landscape and blossoms into acts of kindness and compassion. It's about embracing the path of Karma Yoga, where every action becomes an offering, and every moment becomes an opportunity to serve with humility and love.

In the practice of selfless service, we transcend the narrow confines of the egoic self and expand our awareness to embrace the needs of others. It's not about seeking recognition or reward but about giving freely from the heart, with no expectation of return. Through acts of compassion and kindness, we discover that true fulfillment lies not in the accumulation of wealth or possessions but in the joy of giving and the beauty of selfless service.

As we immerse ourselves in the practice of Karma Yoga, we come to recognize the divine essence that resides within each and every being. We see ourselves reflected in the eyes of others, and we realize that we are all interconnected, all part of the same cosmic tapestry. In serving others with humility and love, we align ourselves with the divine flow of life. We become channels for grace and instruments of peace, spreading light and love wherever we go.

To sum up, cultivating selflessness is a way of living in harmony with the divine essence that dwells within us all.

- **Surrender to the Divine:** Surrendering to the divine is akin to releasing the burdens of the heart and allowing ourselves to be enveloped in the embrace of grace. It's about relinquishing all desires, attachments, and identifications to the divine will, recognizing that the egoic sense of self is a fleeting illusion.

In the act of surrender, we let go of the need to control outcomes and instead trust in the inherent wisdom and grace of the divine to guide our path. We surrender not out of weakness but out of strength, knowing that in letting go, we open ourselves to the infinite possibilities that lie beyond the limited scope of our own understanding.

As we release the grip of the ego, we come to recognize the eternal and unchanging nature of the true Self — the divine essence that resides within us all. We realize that our identity is not confined to the transient fluctuations of the material world but is rooted in the boundless expanse of cosmic consciousness.

Through surrender, we find liberation and discover that true peace and fulfillment lie not in the pursuit of worldly desires but in the surrender to the divine will. When we are aligned with the cosmic order, we find our truest expression of joy and contentment.

- **Seeking Guidance from a Teacher:** Seeking guidance from a spiritual teacher, known as a Guru, is like embarking on a journey with a trusted guide — a wise and compassionate soul who has traversed the path to Moksha and can illuminate our way forward. It's about recognizing that we do not walk this path alone but are supported by the wisdom and guidance of those who have gone before us.

A teacher serves as a torchlight in the darkness, offering support, wisdom, and guidance to the seeker on their quest for liberation. They possess a deep understanding of the spiritual principles and practices that lead to moksha and can help clarify doubts, dispel ignorance, and usher the seeker toward direct experiential realization.

Through the guidance of a Guru, we learn to cultivate humility and surrender, allowing ourselves to be guided by the wisdom of one who has walked the path before us. We open ourselves to the teachings and practices that lead to liberation through this deep and transformative relationship — one that leads us towards the realization of our highest potential and the ultimate goal of human life: liberation from the cycle of birth and death.

- **Cultivating Awareness:** Cultivating awareness is like tending to the garden of our consciousness — a practice that requires gentle nurturing and mindful attention. It's about bringing a sense of presence and mindfulness to every moment of our lives, observing thoughts, emotions, and sensations without getting entangled in them.

  It's easy to get swept away by the currents of our thoughts and emotions in everyday life. That's why, by cultivating awareness, we create a space of inner stillness — a peaceful sanctuary where we can observe the ebb and flow of our inner world with clarity and compassion.

> Through the practice of mindfulness, we learn to remain present and attentive to the here and now, anchoring ourselves in the present moment like a ship amidst a stormy sea. We observe the thoughts that arise in our minds, the emotions that swirl within our hearts, and the sensations that ripple through our bodies, but we do so with a sense of detachment and non-identification.
>
> As we deepen our awareness, we begin to unravel the layers of conditioning and unconscious patterns that shape our perceptions and behaviors. We come to see that our true nature is not defined by the transient fluctuations of the mind but by the eternal aspect of our being — the unchanging essence that lies beyond the realm of thoughts and emotions.
>
> In cultivating awareness, we deepen our connection to the eternal aspect of our being. We come to realize that true peace and fulfillment lie not in the pursuit of external desires but in the cultivation of inner stillness and presence. And as we anchor ourselves in the present moment, we discover a sense of freedom and aliveness that transcends the limitations of the egoic mind.

Essentially, moksha in Vedanta is the crescendo of the soul's symphony — a majestic culmination of the spiritual odyssey. It is the moment when the individual soul awakens to its innate divinity and breaks free from the shackles of samsara, soaring into the boundless expanse of cosmic consciousness.

Through the sacred practices of self-inquiry, meditation, and realization, seekers embark on a sacred pilgrimage towards moksha. They delve deep into the recesses of their own being, uncovering the hidden treasures of their soul and awakening to the eternal flame of divine love that resides within.

In the pursuit of moksha, seekers transcend the illusions of the material world, rising above the dualities of pleasure and pain, good and evil. They move to the cosmic rhythms of existence, embracing the symphony of life with open hearts and minds.

This is a journey of self-discovery, self-realization, and self-transcendence. It is the calling of the soul, the longing for reunion with the divine, and the eternal quest for ultimate truth and liberation.

How does one experience moksha?

You might now be wondering how you know if you have attained moksha. How would that feel and what difference would it make in your life?

Achieving the ultimate liberation in Vedanta philosophy is often described as the realization of one's true nature as divine and eternal, transcending the cycle of birth and death. While the following indicators may provide insight into the state of moksha, it's essential to remember that it is ultimately a subjective and deeply personal realization. Each individual's journey towards moksha is unique, and the experience of liberation may vary from person to person.

Here are some indicators that one may have attained moksha:

Freedom from Suffering: A moksha-realized individual experiences a profound sense of inner peace, contentment, and freedom from suffering. They no longer identify with the fluctuations of the mind or the ups and downs of worldly existence.

Unconditional Love and Compassion: Moksha entails a deep recognition of the interconnectedness of all beings and the expression of unconditional love and compassion towards others. The realized soul sees the divine essence in everyone and acts with kindness and empathy.

Steadiness of Mind: The mind of a moksha-realized being is steady and tranquil, unaffected by external circumstances or inner disturbances. They remain rooted in the awareness of their true nature as pure consciousness, undisturbed by the play of the phenomenal world.

Transcendence of Ego: Moksha involves the transcendence of the egoic sense of self and the realization of one's identity with the ultimate reality (Brahman). The realized soul no longer identifies with the limited ego but abides in the awareness of the infinite and eternal Self.

Liberation from Samsara: Moksha liberates the individual from the cycle of birth and death (Samsara) and the bondage of karma. The realized soul is free from the compulsion of reincarnation and merges with the ultimate reality, experiencing oneness with Brahman.

Effortless Spontaneity: Actions performed by a moksha-realized being arise spontaneously from a place of pure awareness and divine inspiration. There is no sense of doership or attachment to the fruits of actions, as they are guided by the divine will.

Joyful Abiding: Moksha is accompanied by a sense of profound joy, bliss, and fulfillment that transcends the fleeting pleasures of the world. The realized soul abides in a state of eternal happiness, regardless of external circumstances.

Absence of Desire: A moksha-realized individual is free from desires and cravings, as they have realized the ultimate source of fulfillment within themselves. They no longer seek satisfaction from external sources but abide in the bliss of their own being.

Unity of Consciousness: Moksha involves the realization of the underlying unity and interconnectedness of all existence. The realized soul perceives the divine presence in everything and everyone, experiencing a sense of oneness with all creation.

Living in the Eternal Now: Moksha-realized beings live in the eternal present moment, free from regrets of the past or anxieties about the future. They abide in the timeless awareness of their true nature, experiencing the eternal now.

That may seem like a tall order, especially if you are just getting acquainted with Vedanta or starting out on your spiritual path. Remember that there are no set rules and the path to attaining moksha varies for each individual. It depends on various factors such as someone's level of spiritual evolution, intensity of spiritual practice, and degree of purification of mind and heart.

Some individuals may attain moksha in a single lifetime through intense spiritual striving and grace, while others may require multiple lifetimes of spiritual effort and gradual evolution.

Ultimately, the journey towards moksha is timeless and unfolds according to the divine will, with each step bringing individuals closer to the realization of their true nature as divine and eternal.

There are, of course, certain obstacles that need to be overcome for anyone to reach a state of ultimate liberation in Vedanta, such as:

> **Attachment to Worldly Desires:** One of the primary obstacles to moksha is the attachment to worldly desires and the egoic identification with the transient aspects of existence. The pursuit of material pleasures and the accumulation of worldly possessions can distract individuals from seeking higher spiritual truths.
>
> **Ignorance and Illusion:** Moksha requires transcending the veil of ignorance (avidya) and overcoming the illusion (maya) that veils the true nature of reality. Many individuals remain caught in the cycle of samsara due to their mistaken identification with the body, mind, and ego, failing to recognize the eternal and unchanging Self within.

Karmic Bondage: The law of karma dictates that every action has consequences, which can bind individuals to the cycle of birth and death (samsara). Accumulated karma from past actions influences future births and experiences, making it challenging to break free from the cycle of samsara and attain moksha.

Spiritual Ignorance: Lack of spiritual knowledge and guidance can hinder the path to moksha. Without a proper understanding of Vedanta teachings and the practices necessary for self-realization, individuals may struggle to navigate the complexities of the spiritual journey.

Samskaras and Vasanas: Deep-rooted mental impressions (samskaras) and subconscious desires (vasanas) accumulated over countless lifetimes can perpetuate the cycle of samsara and hinder progress toward moksha. Overcoming these ingrained patterns of thought and behavior requires diligent self-inquiry and inner purification.

Egoic Resistance: The ego, or the sense of individuality and separateness, often resists the dissolution required for moksha. The ego clings to its identity and fears annihilation, making it difficult for individuals to surrender fully to the divine and merge with the ultimate reality.

As we conclude this chapter, let us be reminded of the profound journey toward enlightenment — a journey that tests the limits of our resilience, courage, and wisdom. Despite the challenges and obstacles that may arise, it is through facing these trials that we uncover the depths of our inner strength and fortitude.

The path to enlightenment demands dedication, patience, and humility as we navigate the complexities of the mind and explore the mysteries of existence. Yet, with each step forward, we draw closer to the radiant dawn of self-realization, where the light of divine truth shines brightly, illuminating our path with boundless grace and wisdom.

May we continue our journey with unwavering resolve, knowing that beyond the trials and tribulations lies the eternal bliss of moksha — a realm where the soul finds ultimate liberation and transcendence.

And I am free

Bhante, I want to be free.

What does that mean to you?

I want to be free from suffering.

And how do you think it's possible?

If I meditate and clean my karma.

And how do you achieve that?

By having a daily practice.

Good. So what's in the way?

I don't have enough time and discipline.

How can you start?

By having the freedom to do what is truly needed in life.

Say the same thing slowly again to yourself and this time, notice what comes up.

I guess I am the source of my own freedom and liberation.

Yes! Truly experiencing that, albeit at the level of your soul, is the true moksha.

Thank you, Bhante.

You are welcome. Now is the time!

"My actions are my only true belongings. I cannot escape the consequences of my actions. My actions are the ground upon which I stand."

– Thich Nhat Hanh

CHAPTER 14

KARMA: THE CAUSAL CONNECTION

If I had a dollar for every time I heard the word' karma' being misused, I would most likely be a multimillionaire, if not a billionaire by now. That being said, karma is perhaps the only word from the Vedas that beats *Tantra* on the list of most misunderstood concepts, having taken on a somewhat skewed meaning in popular culture since first being introduced to the West in the 19th century.

Today, there are times when I literally cringe at the word being used in some contexts – for example, on an Instagram meme declaring that '*karma is a b*tch',* as if it is some vindictive entity waging retribution on all who do wrong. That's far-removed from its original meaning in the Eastern traditions and one I wish to attempt to rectify here.

Using the word karma to mean "the results of past actions" inherited from previous lives, and exclaiming, "It's my karma," or "It's bad karma," to convey a sense of fatalism and luck does not align with the source meaning of the word at all.

In fact, the Sanskrit root of the word "karma" (कर्म) means "action" or "deed" and has nothing to do with fate or destiny. Karma is both one's actions and their consequences because every action leaves an imprint (in a transcendent sense of the word), but it isn't some mysterious force we have no control over. I'll get into the intricacies of what karma really means a bit later on but want to first take a look at how the notion of karma has become so distorted in the West.

The evolution of understanding/misunderstanding in the West:

The concept of karma, originating in Eastern civilizations, began to permeate Western thought in the 19th and 20th centuries through the efforts of influential figures like Ralph Waldo Emerson, Henry David Thoreau, and Swami Vivekananda. These thinkers introduced and interpreted karma within Western intellectual and spiritual circles, shaping contemporary understanding. However, cultural differences and popularization through media have led to varied interpretations and misunderstandings. In Western popular culture, karma is often simplified to a cause-and-effect system, overlooking its deeper spiritual and philosophical dimensions. Furthermore, New Age movements have appropriated karma, diluting its traditional meanings and emphasizing simplistic reward-and-punishment dynamics.

In addition, Western emphasis on individualism has shaped the understanding of karma, resulting in interpretations focused on personal actions and outcomes. Ideas such as "karmic accountability" encourage self-empowerment but may neglect karma's spiritual depth and interconnectedness. Overall, the Western view of karma may prioritize individual agency over broader cosmic or spiritual principles found in Eastern traditions.

Various interpretations of karma have also emerged due to the diverse religious and philosophical beliefs embraced by the West. Some may view it through a religious lens, while others approach it from a secular or philosophical perspective, leading to different understandings and applications.

Imagine a group of individuals from different religious and philosophical backgrounds gathering for a panel discussion on the concept of karma in a multicultural society. Among the panelists are a Hindu priest, a Buddhist scholar, a secular philosopher, and a New Age practitioner. Each panelist offers their perspective on karma based on their respective beliefs and worldviews.

The Hindu priest might explain karma as a fundamental principle of Hinduism, emphasizing its role in shaping individual experiences and influencing the cycle of birth and rebirth. He will discuss the concept of dharma (duty) and the importance of performing righteous actions to accumulate positive karma and progress spiritually.

The Buddhist scholar will interpret karma within the context of Buddhist teachings, highlighting its role in understanding suffering and the path to enlightenment. They will discuss the concept of dependent origination and how karma operates within the framework of the Four Noble Truths and the Noble Eightfold Path.

The secular philosopher will approach karma from a philosophical standpoint, exploring its implications for ethics, morality, and human agency. They will discuss different philosophical theories of karma, including determinism, compatibilism, and libertarianism, and engage in a debate on the nature of free will and moral responsibility.

The New Age practitioner will offer a more eclectic interpretation of karma, drawing from various spiritual traditions and metaphysical beliefs. They will discuss karma in the context of energy dynamics, vibrational frequencies, and the law of attraction, suggesting that individuals can influence their karma through positive thoughts, intentions, and actions.

Each panelist will approach karma from a different perspective, reflecting the pluralistic nature of Western society. While some will view karma through a religious lens, others will approach it from a secular or philosophical standpoint, leading to different understandings and applications of the concept. Nonetheless, this diversity enriches the discourse on karma, fostering dialogue and exchange of ideas across different traditions and perspectives.

Overall, we can see how the misunderstanding of karma in the West can be attributed to a combination of factors: cultural differences, oversimplification, appropriation, and selective adoption, which have shaped how the concept is perceived and interpreted outside of its original cultural and religious contexts.

What, then, is Karma?

The concept of karma first appears in ancient Indian texts such as the Vedas and the Upanishads. However, the precise origins of the concept are difficult to pinpoint, as it evolved over centuries within the rich tapestry of Indian thought.

In Vedanta, karma is a fundamental principle that governs the cycle of birth, death, and rebirth (samsara) and is intricately linked to the concept of dharma (duty, righteousness). The earliest references to karma in Hindu scriptures can be found in the Brahmanas and the Upanishads, where it is discussed in the context of ritual action, moral conduct, and cosmic order. You will find some relevant quotes for each source below:

- From the Brahmanas:

  "As one acts, so does he become. One becomes virtuous by virtuous action, bad by bad action." (Shatapatha Brahmana 10.1.1.8)

  This passage emphasizes the idea that individuals shape their own destinies through their actions. It reflects the concept of karma as a law of cause and effect, where virtuous actions lead to positive outcomes, while sinful actions result in negative consequences. This principle underscores the importance of moral conduct and the accountability of individuals for their deeds.

  "Whatever deed he does — good or bad — of that he shall be the heir." (Taittiriya Brahmana 3.11.8.6)

  This statement reinforces the notion that individuals inherit the consequences of their actions, whether they are virtuous or sinful. It suggests that one's deeds determine their future experiences and shape their destiny. This aligns with the concept of karma as a moral law governing the cycle of cause and effect, where individuals reap the fruits of their actions in this life or in subsequent existences.

- From the Upanishads:

  "By the past deeds is here caused the condition of beings; by the future, that will be caused by

> *deeds; by the present, they act in the world." (Chandogya Upanishad 5.24.1)*
>
> This passage highlights the temporal dimensions of karma, suggesting that past actions influence present circumstances, and present actions shape future outcomes. It reflects the cyclic nature of karma, where individuals are bound by their past deeds ***(prarabdha karma)*** but also have the opportunity to create new karma through their present actions. This underscores the dynamic interplay between past, present, and future in the operation of karmic law.
>
> *"According as one acts, so does he become. One becomes virtuous by virtuous action, bad by bad action." (Brihadaranyaka Upanishad 4.4.5)*
>
> This passage echoes similar sentiments found in the Brahmanas, emphasizing the connection between action and character. It reinforces the idea that individuals shape their destinies through their actions, with virtuous deeds leading to spiritual growth and evolution, while sinful deeds result in moral degradation. This aligns with the core principle of karma as a law of cause and effect, where individuals are responsible for their own karma and its consequences.

These quotes provide insights into the ancient Indian understanding of action and its consequences, which form the basis of the concept of karma in Hindu philosophy.

They emphasize the moral and ethical dimensions of karma, underscoring the importance of righteous conduct and the accountability of individuals for their deeds.

In Buddhism, karma is a central concept that holds significant importance in understanding the nature of existence and the path to liberation. While the basic premise of karma as intentional action and its consequences is shared with Hinduism, Buddhism offers a unique perspective on karma and its implications. Here, karma is understood as intentional actions (kamma in Pali) that have the power to shape one's future experiences and determine one's destiny within the cycle of rebirth (samsara). Every action, whether physical, verbal, or mental, leaves an imprint on the individual's consciousness (karmic imprint), which influences their subsequent lives and experiences.

This understanding of karma is intricately linked to the Buddhist concept of dependent origination (paticca samuppada), which elucidates the interconnectedness of phenomena and the causes of suffering.

The concept of karma in Buddhism plays a crucial role in elucidating the workings of cause and effect, illustrating how one's actions give rise to various states of existence and the perpetuation of samsara. Through the understanding of karma, Buddhists recognize the role of volition (chetana) in shaping their destinies and the importance of cultivating wholesome intentions and actions to alleviate suffering and attain liberation.

Moreover, karma is closely tied to the Buddhist path to liberation (nirvana), which aims to break free from the cycle of birth and death. By purifying one's intentions and actions, practicing ethical conduct (sila), cultivating mental discipline (samadhi), and developing wisdom (panna), individuals can gradually weaken the hold of karma and ultimately attain liberation from the cycle of samsara. Thus, karma serves as a guiding principle on the path to enlightenment, reminding practitioners of the ethical responsibilities inherent in their actions and the potential for spiritual growth and transformation.

Similarly, in Jainism and Sikhism, karma is a fundamental concept that influences moral behavior, spiritual progress, and the cycle of birth and death. In Jainism, karma is intricately woven into the fabric of the universe, governing the cycle of birth, death, and rebirth (samsara) as well as the individual journey towards spiritual liberation (moksha). Jains believe in the existence of countless souls (jivas), each of which is bound by karma accumulated through their thoughts, words, and deeds. The concept of karma in Jainism is highly nuanced, with emphasis placed on the intention behind actions and the subtleties of karma particles (karma pudgala) that bind the soul to the cycle of transmigration. Jain teachings advocate for the practice of non-violence (ahimsa), truthfulness (satya), and other virtues as a means of minimizing the accrual of karma and attaining spiritual purity.

Meanwhile, in Sikhism, karma plays a crucial role in shaping individual behavior and spiritual evolution. Sikhs believe in the concept of karmic accountability, wherein every action generates consequences that affect one's present life and future incarnations. However, Sikhism emphasizes the importance of surrendering to the divine will (hukam) and cultivating a relationship with the

Supreme Being (Waheguru) as a means of transcending the cycle of karma and achieving spiritual liberation. Sikhs are encouraged to live virtuously, serve others selflessly, and meditate on the divine name (Naam Simran) to purify their consciousness and align themselves with the divine order.

Overall, the concept of karma is deeply rooted in ancient Indian religious and philosophical tradition yet it has more than one interpretation. Interpretation. It appears in various forms across different religious and philosophical systems of the Indian subcontinent.

Karma in Vedanta

In Vedanta, the concept of karma is deeply intertwined with the understanding of the self (atman), the ultimate reality (Brahman), and the nature of existence. Since Vedanta emphasizes the pursuit of self-realization and liberation (moksha) from the cycle of birth and death (samsara), karma plays a crucial role by shaping individual experiences and contributing to spiritual evolution.

Let's take a look below at some key aspects of how the concept of karma is understood in Vedanta.

1. **Law of Cause and Effect:**

   Vedanta views karma as the law of cause and effect, whereby every action (karma) produces corresponding results (phala). This principle underscores the interconnectedness of all beings and the moral responsibility inherent in one's actions. Actions can be physical, verbal, or mental, and each action generates a karmic imprint that influences future experiences.

Vedanta views karma as a universal law that governs the consequences of one's actions. Just as in Newton's third law of motion, "For every action, there is an equal and opposite reaction," Vedanta asserts that every action, whether physical, verbal, or mental, generates a corresponding result. This principle of cause and effect highlights the interconnectedness of all beings and events in the universe, illustrating how actions reverberate through time and space, shaping individual destinies and the collective human experience.

This understanding is reflected in various Vedantic texts, including the Bhagavad Gita (3.5), where Lord Krishna states, "*One must perform one's duty, following the ways of one's own nature; one should not incur sin.*"

Another quote from the same source, which is personally one of my greatest inspirations, states:

कर्मण्येवाधिकारस्ते मा फलेषु कदाचन।
मा कर्मफलहेतुर्भूर्मा ते सङ्गोऽस्त्वकर्मणि॥ २-४७

"You have a right to perform your prescribed duties, but you are not entitled to the fruits of your actions. Never consider yourself to be the cause of the results of your activities, nor be attached to inaction." (Bhagavad Gita 2.47)

In other words, individuals have a duty to perform their prescribed actions (karma), but they should not be attached to the outcomes (phala) of those actions. This also highlights the illusion of attachment to "inaction" waiting for the fruits.

This principle underscores the concept of karma as the law of cause and effect, where actions generate consequences but individuals are not the ultimate controllers of those outcomes. The passage highlights the importance of performing one's duties selflessly, without attachment or ego, and accepting whatever results arise as part of the cosmic order. This verse serves as a foundational teaching on karma within the Bhagavad Gita and is widely cited in discussions on Vedanta philosophy.

Vedanta recognizes that actions encompass not only physical deeds but also speech and thoughts. Every action, whether overt or subtle, leaves an imprint on the individual's consciousness, known as a karmic seed or impression (samskara), which influences future experiences and tendencies. Thus, one's karma is not limited to external actions but also includes the intentions and attitudes underlying those actions. This understanding underscores the moral responsibility inherent in one's thoughts, words, and deeds, emphasizing the importance of cultivating virtuous qualities and mindfulness in all aspects of life.

2. **Accumulation of Karma:**

According to Vedanta, individuals accumulate karma over successive lifetimes as a result of their actions, intentions, desires, and attitudes.

These karmic impressions (samskaras) shape the circumstances of one's life and contribute to the cycle of birth and death.

"As you sow, so shall you reap."

This proverbial saying succinctly captures the essence of the accumulation of karma in Vedanta philosophy. It implies that the actions (karma) one performs determine the outcomes or consequences (phala) one experiences. This concept underscores the idea that individuals accumulate karma over successive lifetimes through their actions, intentions, desires, and attitudes.

"The soul acquires merit by good actions and demerit by evil actions. It is with the merit and demerit acquired by itself that it passes from womb to womb." (Chandogya Upanishad 5.10.7)

This quote elucidates how individuals accumulate karma and its consequences over multiple lifetimes. It suggests that the soul (atman) undergoes the cycle of birth and rebirth (samsara) based on the merits and demerits acquired through its own actions. This process highlights the karmic responsibility inherent in individual conduct and its role in shaping one's destiny across the continuum of existence.

"Desire for wealth, longing for pleasures, are the fruits of actions done in the past." (Brihadaranyaka Upanishad 4.4.5)

Here, we learn how desires and inclinations in the present are influenced by past actions and karmic impressions. It suggests that individuals experience the fruits of their past actions in the form of desires and cravings that manifest in their current lives. This understanding underscores the intricate relationship between past karma, present experiences, and future destinies in the cycle of samsara.

3. **Three Types of Karma**:

Vedanta classifies karma into three categories based on their effects: *Sanchita karma (accumulated karma)*, *Prarabdha karma (fructifying karma)*, and *Agami karma (future karma)*. Sanchita karma refers to the sum total of all past actions, Prarabdha karma refers to the portion of Sanchita karma that is currently unfolding in one's life, and Agami karma refers to actions performed in the present that will yield future results.

I. Sanchita Karma (Accumulated Karma):

"Sanchita karma is the accumulated karma of all past actions."

Sanchita karma refers to the sum total of all actions (karma) performed in past lives that have not yet borne fruit. It represents the storehouse of karmic impressions (samskaras) accumulated over successive lifetimes. This reservoir of karma influences the circumstances of one's current life and future incarnations, shaping the experiences and destinies of individuals within the cycle of samsara.

II. Prarabdha Karma (Fructifying Karma):

"Prarabdha karma is the portion of Sanchita karma that is currently unfolding in one's life."

Prarabdha karma refers to the subset of Sanchita karma that has ripened and is manifesting as the experiences and situations encountered in one's present life. It represents the predetermined destiny or life script that individuals are destined to experience in their current incarnation. Prarabdha karma determines various aspects of life, such as birth, family, social status, and major life events, serving as the catalyst for spiritual growth and karmic resolution.

III. Agami Karma (Future Karma):

"Agami karma refers to actions performed in the present that will yield future results."

Agami karma pertains to the actions (karma) performed in the present moment, which will accrue and shape future experiences and destinies. It represents the ongoing accumulation of new karma through thoughts, words, and deeds, influencing the trajectory of one's journey within the cycle of samsara. Agami karma underscores the significance of mindful action and ethical conduct in creating positive outcomes and mitigating negative consequences in future lives.

The classification of karma into these three categories provides a framework for understanding the dynamics of cause and effect (karma) and its implications for individual spiritual evolution. By recognizing the interplay between past, present, and future karma, individuals gain insight into the complexities of their karmic journey and the opportunities for growth and liberation within the cosmic order. This understanding encourages practitioners to cultivate awareness, discernment, and virtuous conduct in navigating the intricacies of karma and progressing towards ultimate liberation (moksha).

4. **Karma and Free Will:**

Vedanta acknowledges the role of free will in shaping one's actions but also recognizes the influence of past karma on present circumstances. While individuals have the freedom to choose their actions, they are also subject to the consequences of their past actions, which may limit their choices and experiences.

"Knowledge, power, and action are all in accordance with one's nature." (Mundaka Upanishad 3.1.6)

This text elucidates the idea that individuals act in accordance with their inherent nature (swabhava). While individuals possess free will and the capacity to make choices, their actions are influenced by their inherent tendencies, inclinations, and past karma. It suggests that while individuals have the freedom to act, their choices are shaped by their past experiences, karmic predispositions, and the interplay of various factors. We read of the intricate relationship between individual agency and karmic conditioning in determining one's actions and experiences.

"For, the one who has control over the mind, and is ever absorbed in the Self, one who is free from desires, experiences the bliss of the Self, and does not encounter sorrow from anything; for, it is the mind, O Arjuna, that is the cause of bondage, as well as liberation!" (Bhagavad Gita 6.7)

This verse from the Bhagavad Gita highlights the importance of mastering the mind and cultivating inner freedom. It suggests that individuals who have control over their minds and are absorbed in the realization of the Self (Atman) experience a state of inner bliss and freedom from suffering. The verse implies that true liberation (moksha) comes from transcending desires and attachments, which are often the result of past karma. By gaining mastery over the mind and aligning oneself with the higher Self, individuals can exercise their free will to break free from the cycle of karma and attain ultimate liberation. This verse emphasizes the transformative power of self-awareness, mindfulness, and spiritual realization in transcending the influence of past karma and realizing one's innate freedom and bliss.

The relationship between karma and free will is characterized by a dynamic interplay between individual agency and the influence of past actions. This understanding encourages individuals to exercise discernment, ethical conduct, and mindfulness in navigating the complexities of karma and realizing their potential for spiritual growth and liberation.

5. **Purification and Liberation:**

The ultimate goal of karma in Vedanta is not merely to accumulate positive karma for worldly gains but to purify the mind, transcend egoic desires, and attain liberation (moksha) from the cycle of birth and death. By performing selfless actions (karma yoga), cultivating knowledge (jnana yoga), and devotion (bhakti yoga), individuals can transcend the limitations of karma and realize their true nature as the eternal Self (atman) united with Brahman.

The ultimate goal of karma in Vedanta:

"You have a right to perform your prescribed duties, but you are not entitled to the fruits of your actions. Never consider yourself to be the cause of the results of your activities, nor be attached to inaction." (Bhagavad Gita 2.47)

This verse underscores the principle of karma yoga, where individuals are encouraged to perform their duties selflessly, without attachment to the outcomes. By relinquishing attachment to the results of their actions, individuals can purify their minds and transcend egoic desires, ultimately leading to liberation from the cycle of birth and death.

Performing selfless actions (karma yoga):

"Therefore, without being attached to the fruits of activities, one should act as a matter of duty;

for by working without attachment, one attains the Supreme." (Bhagavad Gita 3.19)

Lord Krishna advises Arjuna to perform his duties without attachment to the results. By acting selflessly and without desire for personal gain, individuals can attain the highest fulfillment and liberation from the cycle of samsara. This teaching emphasizes the importance of performing actions as a service to the higher Self, leading to purification of the mind and spiritual growth.

Cultivating knowledge (jnana yoga):

"Truth alone triumphs; not falsehood. Through truth, the divine path is spread out by which the sages whose desires have been completely fulfilled, reach where that supreme treasure of Truth resides." (Mundaka Upanishad 1.1.3)

This verse extols the significance of truth and knowledge in the pursuit of liberation. It emphasizes that truth alone leads to the highest goal, where seekers realize the supreme truth of their nature and attain liberation from the cycle of birth and death. This teaching underscores the importance of spiritual wisdom (jnana) in dispelling ignorance and realizing the true nature of the Self, leading to ultimate freedom.

Devotion (bhakti yoga):

"Engage your mind always in thinking of Me, offer obeisances and worship Me. Being completely absorbed in Me, surely you will come to Me." (Bhagavad Gita 9.3)

In this verse, Lord Krishna instructs devotees to constantly focus their minds on Him, offer worship, and surrender to Him with love and devotion. By practicing devotion to the Divine with sincerity and devotion, individuals can realize their inherent unity with the Supreme and attain liberation from the cycle of samsara. This teaching highlights the path of bhakti yoga, emphasizing the importance of cultivating a loving relationship with the Divine as a means to spiritual realization and ultimate liberation.

These quotes serve as foundational teachings in Vedanta philosophy, elucidating the paths of karma yoga, jnana yoga, and bhakti yoga as means to purification and liberation from the cycle of birth and death. Through selfless actions, spiritual knowledge, and devotion to the Divine, individuals can transcend the limitations of karma and realize their true nature as the eternal Self united with Brahman.

Karma: The Cosmic Connection

Karma extends its influence far beyond individual actions, reaching into the vastness of the universe and interconnecting with the movements of stars and planets.

Vedic cosmology encompasses a rich understanding of the universe, where karma plays a central role in shaping the cosmic order. The cosmos is seen as a dynamic, interconnected web of energy and consciousness, governed by divine laws and principles. Karma, in this context, refers to the law of cause and effect that governs the moral and spiritual consequences of one's actions.

According to Vedic cosmology, the universe operates according to a cyclical pattern of creation, preservation, and dissolution, known as the cosmic cycles or "yugas." These yugas are characterized by specific spiritual qualities and evolutionary stages, with each yuga representing a different phase in the cosmic evolution of consciousness.

Karma is intricately woven into the fabric of the cosmos, influencing the destiny of beings across various planes of existence. In the Vedic worldview, every thought, word, and action generates karma, which accumulates over lifetimes and shapes one's future experiences. Positive actions lead to favorable outcomes and spiritual progress, while negative actions result in adverse consequences and karmic debt.

The movement of stars and planets is believed to reflect the cosmic intelligence and divine order governing the universe. Vedic astrology, known as "Jyotish" is based on the idea that the positions of celestial bodies at the time of a person's birth influence their karmic patterns and life experiences. By studying the positions and alignments of planets, astrologers seek to gain insight into the karmic lessons and potentials of individuals.

Furthermore, Vedic rituals and practices are aimed at harmonizing with the cosmic order and mitigating the effects of negative karma. Through acts of selfless service (karma yoga), devotion (bhakti yoga), meditation (dhyana yoga), and self-inquiry (jnana yoga), individuals seek to purify their karma, transcend the cycle of birth and death, and attain spiritual liberation (moksha).

To that extent, we can say that Vedic cosmology provides a holistic framework for understanding the relationship between karma and the universe. Karma is an integral part of the cosmic order, influencing the evolution of consciousness and the destiny of beings within the vast expanse of creation.

How can I embrace Vedanta's concept of karma in my life?

While the concept of Karma may seem more complicated than you originally thought, it really isn't. It may just be different from what you had in mind up until now so may need a little reworking. Karma is simply energy; it's our intentional thoughts and actions. The energy we generate now and in the future will affect us. It has nothing to do with reward or punishment. In essence, Karma is unbiased and it's ours to control.

To better grasp the Vedic concept of karma in relation to your own spirituality, you can take several steps. Firstly, you can study the sacred scriptures of Vedanta, such as the Bhagavad Gita, Upanishads, and Vedas, to gain a deeper understanding of the philosophical underpinnings of karma, its implications for spiritual growth, and its role in the cycle of birth and death.

You might also wish to reflect on personal experiences and how the principle of karma has played out in your journey. Consider those moments of joy, suffering, success, and challenges, and contemplate how your actions and choices may have influenced these outcomes.

Self-reflection is also a useful habit to cultivate as it allows you to examine your thoughts, emotions, and behaviors. By exploring how your intentions, desires, and attitudes shape your actions, you will also discover how they contribute to the accumulation of karma.

At the same time, mindfulness practices such as meditation, yoga, or mindfulness exercises to cultivate present-moment awareness will deepen your understanding of the interconnectedness of all beings. You will notice the ripple effects of your actions and how they impact yourself and others.

Why not embrace the principle of karma as a call to take responsibility for your thoughts, words, and deeds? This will enable you to recognize that you have the power to shape your destiny through conscious choices and intentional actions.

Cultivating compassion and forgiveness towards yourself and others is a large part of the journey as you navigate the complexities of karma. Recognize that everyone is on their own path of growth and transformation, and offer kindness and understanding along the way.

You can always seek guidance from spiritual teachers, mentors, or community members who have a deep understanding of Vedanta and can offer insights and support on your spiritual path. Through meaningful conversations and the exchanging of ideas, you will deepen your understanding of karma and its implications for your spirituality.

On a practical level, Vedanta proposes various ways for you to apply the principles of karma in your life on a daily basis:

1. Embrace the Spirit of Selfless Service:
    - Approach your daily tasks and responsibilities with a sense of purpose and dedication, focusing on the process rather than the outcome.
    - Look for opportunities to serve others without expecting anything in return, whether it's lending a helping hand to a neighbor or volunteering for a cause you believe in.
    - Cultivate a mindset of generosity and compassion, seeking to make a positive difference in the lives of those around you through small acts of kindness and consideration.
2. Cultivate a Heart of Devotion:
    - Establish a daily ritual that connects you with the Divine, whether it's through prayer, meditation, chanting, or simply spending quiet time in nature.
    - Infuse your daily activities with a sense of reverence and gratitude, recognizing the sacredness of life and the interconnected web of existence.
    - Express your devotion through acts of kindness and service, seeing the Divine in all beings and treating them with respect, compassion, and love.
3. Embrace Life with Acceptance and Equanimity:
    - Embrace life's inevitable ups and downs with an attitude of acceptance and equanimity,

recognizing that every experience holds valuable lessons and opportunities for growth.

- Cultivate resilience and inner strength by facing challenges with courage and grace, trusting in the wisdom of the universe and your ability to navigate through adversity.
- Practice letting go of attachments and expectations, allowing life to unfold naturally while remaining grounded in the present moment and open to the possibilities that arise.

These practical tips can infuse each moment with meaning, purpose, and spiritual depth, ultimately aligning your actions with the timeless wisdom of Vedanta and journeying toward greater fulfillment, self-realization, and inner peace.

As we conclude this enlightening chapter on Karma, we are reminded of the profound wisdom and timeless truths embedded within Vedanta. Through exploring the intricate motion of cause and effect, we have discovered the importance of our actions and intentions.

Vedanta teaches us that karma is not merely a cosmic ledger of deeds, but a profound spiritual principle that guides our journey towards self-realization and liberation. As we navigate the complexities of life, the teachings of karma help us recognize the interconnectedness of all beings and the boundless potential for growth and transformation.

Let us walk this path with integrity, compassion, and wisdom, knowing that every thought, word, and action has the power to shape our destiny and contribute to the evolution of consciousness. With this understanding, may you embark on our spiritual journey with courage, curiosity, and an open heart, ever seeking to awaken to the eternal truth that lies within.

When Karma moves silently

In the bustling streets of Chicago, Maya stumbled upon a lost dog while out for a jog along Lake Shore Drive. She had been in that part of Chicago to visit the family of her best friend from boarding school in Virginia. Maya found herself torn between rushing to a long-awaited date with her favorite friend, whom she had a secret crush on, or attempting to look for the dog's owner. Despite her desire to go on her date, Maya chose to prioritize finding the lost dog's owner.

She spent hours combing through the nearby neighborhoods, determined to reunite the furry wanderer with its owner. Eventually, aided by social media and some regular joggers, her search led her to a charming brownstone in the historic Lincoln Park area.

Grateful for Maya's efforts, the owner, a renowned chef named Antonio, invited her to the grand opening of his new restaurant, nestled amidst the vibrant culinary scene of the Windy City. Excited by the invitation, Maya eagerly marked the event on her calendar. However, fate had other plans. On the day of the grand opening, her grandmother was sick and she could not go to the event.

Years later, Maya returned to Chicago to pursue an internship in culinary arts at one of the city's prestigious culinary schools. It was there, in the heart of the city's bustling food scene, that she crossed paths with Michael, a fellow student with a passion for cooking that rivaled her own.

As they exchanged stories in their after-classroom socializing, Maya discovered that Michael was Antonio's son — the very person whose lost dog she had found years before. With a smile, Michael recounted how Maya's kindness had not only reunited him with his beloved pet but also left an indelible mark on his heart.

Amidst the hustle and bustle of the city, Maya and Michael's bond blossomed, fueled by their shared love for food and the serendipitous connection forged by Maya's act of kindness all those years ago. And as they navigated the busy streets of Chicago together, Maya couldn't help but marvel at the intricate tapestry of fate that had brought them together in the city they both called home.

The dog, incidentally, had been adopted from a rescue center as a puppy by Michael's mother, Alejandra. At the time of adoption, she had affectionately named him *Karma*.

This story was narrated to me by Michael's sister who was my batchmate in business school. She was suitably amused as I explained to her what "Karma" means to Hindus. Maya and Michael are now married and still live in Chicago!

"It was only my first meeting with the Vedas, but I seemed to have heard echoes of their sublime teachings all my life."

– W.B. Yeats

PART IV:

SPECIALIST PATHS IN VEDANTA

Vedanta is a rich synthesis of diverse and varied paths — some more well-known than the others, some more misunderstood than the others. In this section, we will discuss two specialist paths within Vedanta that are probably both lesser known and also somewhat misunderstood. So with that briefest of introductions, let's embark on a journey through the paths of Brahma Vidya and Tantra.

We should begin by acknowledging that, while Jnana Yoga, Bhakti Yoga, Karma Yoga, and Raja Yoga are widely recognized, Brahma Vidya and Tantra offer deeper and more specialized avenues for spiritual exploration within Vedanta.

Brahma Vidya, or the Science of Brahman, delves into profound philosophical inquiry, contemplation, and meditation to grasp the essence of Brahman, the ultimate reality. Practitioners engage in the rigorous study of foundational texts like the Upanishads and the Bhagavad Gita, guided by commentaries from revered sages and scholars. Through deep introspection and meditation, they seek to unravel the mysteries of existence and realize the non-dual nature of reality.

On the other hand, Tantra encompasses a diverse array of practices that aim to expand consciousness and awaken spiritual energy. These practices may include ritual worship, mantra recitation, and visualization techniques, often accompanied by the study of Tantric texts and the guidance of experienced mentors. There are also parts within this tradition that emphasize the synthesis of sexual energy and channelizing it for spiritual growth — a topic that is little understood and quite controversial in some ways. Overall, Tantra offers a path to spiritual awakening through direct experience and transformation, embracing the sacredness of everyday life and the interconnectedness of all things.

Why specialized paths?

As the rich tapestry of Vedanta unfolds before us, we uncover a myriad of paths tailored to suit the diverse needs and yearnings of spiritual seekers on their quest for self-realization and liberation. Vedanta acknowledges that each individual walks a unique path, shaped by their temperament, cultural background, and spiritual evolution.

Imagine navigating a vast landscape of spiritual possibilities, where every trail offers a different vista, a different experience. For those inclined towards intellectual inquiry and deep contemplation, Jnana Yoga beckons with its profound philosophical teachings. Others may find solace in the heart-opening practices of Bhakti Yoga, surrendering to the divine with love and devotion.

Cultural influences play a significant role in shaping our spiritual orientation. Whether rooted in devotional traditions or scholarly pursuits, Vedanta accommodates these cultural variations with specialized paths that resonate with seekers from all walks of life.

And let's not forget about timing — where you are on your spiritual journey matters. Vedanta honors the diversity of seekers, offering practices and teachings suited to their level of readiness and spiritual maturity. Whether you're taking your first tentative steps or you're already immersed in the depths of self-realization, there's a path tailored just for you.

Moreover, Vedanta embraces the idea that all paths, no matter how diverse, ultimately lead to the same destination – self-realization and liberation. This inclusive outlook fosters a sense of unity amidst diversity, honoring the multiplicity of paths leading to the ultimate truth.

But traversing these paths isn't a solitary journey. Seekers often form close bonds with spiritual teachers (gurus), who offer personalized guidance and support along the way. This intimate relationship allows for deeper exploration and understanding, paving the way for profound spiritual insights and transformation.

Primary sources

The paths within Vedanta draw upon a variety of sources, some of which we have already looked at in previous chapters. They also include ancient scriptures, philosophical treatises, commentaries by revered saints and sages, and the oral tradition passed down through generations of spiritual teachers (gurus).

The primary sources that inform these specialized paths are:

Upanishads: As we now know, the Upanishads are considered to be the foundational texts of Vedanta philosophy. They contain profound insights into the nature of reality, the self (Atman), and the ultimate goal of life (moksha). Different Upanishads emphasize various aspects of spiritual practice and philosophical inquiry, serving as sources of inspiration and guidance for specialized paths within Vedanta.

Bhagavad Gita: The Bhagavad Gita, a sacred dialogue between Lord Krishna and the warrior prince Arjuna, expounds on various paths to spiritual realization. It discussed the concepts of Karma Yoga (the path of selfless action), Bhakti Yoga (the path of devotion), and Jnana Yoga (the path of knowledge). Its teachings provide valuable insights into the practical application of Vedanta philosophy in everyday life, guiding seekers on their spiritual journey.

Brahmasutras: Also known as the Vedanta Sutras, the Brahmasutras are aphoristic texts attributed to the sage Vyasa. They systematically elucidate the teachings of the Upanishads and provide a framework for understanding and interpreting Vedanta philosophy. The

Brahmasutras serve as a foundational text for specialized paths within Vedanta.

Vedantic Commentaries: Throughout history, numerous scholars and spiritual luminaries have written commentaries on the Upanishads, Bhagavad Gita, and Brahmasutras, offering deeper insights and interpretations of Vedanta philosophy. These commentaries serve as authoritative sources for understanding specialized paths within Vedanta.

Guru Parampara: The oral tradition passed down from spiritual teacher to disciple, known as the guru parampara, is a vital source of knowledge and guidance in Vedanta. Enlightened masters impart teachings, practices, and insights based on their own direct experience of truth, guiding seekers on the path to self-realization. The teachings of realized gurus are considered invaluable sources of wisdom and inspiration for practitioners of specialized paths within Vedanta.

Ancient Texts and Scriptures: In addition to the core Vedantic texts, specialized paths within Vedanta may draw inspiration from a wide range of ancillary texts, scriptures, and philosophical works from various spiritual traditions. These texts provide diverse perspectives on spiritual practice and philosophical inquiry, enriching the tapestry of Vedanta philosophy and informing specialized paths tailored to individual needs and inclinations.

Overall, the sources within Vedanta are diverse and multifaceted, encompassing a rich mosaic of ancient scriptures, philosophical treatises, commentaries, oral traditions, and the wisdom imparted by enlightened masters throughout history.

Let's take a closer look at two of these highly specialized paths below: Brahma Vidya and Tantra.

"The religion of the future will be a cosmic religion. It should transcend personal God and avoid dogma and theology. Covering both the natural and the spiritual, it should be based on a religious sense arising from the experience of all things natural and spiritual as a meaningful unity."

– Albert Einstein

CHAPTER 15

BRAHMA VIDYA: THE SCIENCE OF ULTIMATE REALITY

The first path I want to talk to you about is Brahma Vidya. It is also known as Brahmavidya or the Science of Brahman. More rigorously speaking, it is the 'study and knowledge of the Brahman". Imagine a path less traveled, shrouded in the mysteries of the cosmos and beckoning seekers to unravel the enigma of existence. This is Brahma Vidya; a sacred journey into the heart of Vedanta philosophy, where seekers embark on a quest for the direct realization of Brahman, the supreme reality.

More than just a pursuit of knowledge, Brahma Vidya is an immersive exploration of the very essence of existence. The term itself hints at the profundity of this path, translating to "knowledge of Brahman" or "the science of the absolute." Here, seekers delve deep into the intricate web of Vedantic teachings, probing the fundamental nature of Brahman and the individual self (Atman) with relentless curiosity.

But Brahma Vidya isn't for the faint of heart. It demands courage, patience, and unwavering dedication as seekers navigate the labyrinth of metaphysical concepts and philosophical inquiries. They question the nature of reality itself, peeling away layers of illusion to reveal the timeless truth that lies beneath.

At the heart of Brahma Vidya lies a transformative journey of self-discovery and inner exploration. Seekers venture beyond the realm of ordinary perception, transcending the limitations of the mind and intellect to directly experience the sublime reality of Brahman. Through rigorous practices like meditation, self-inquiry, and introspection, they strive to silence the restless chatter of the mind and tap into the infinite reservoir of consciousness within.

This journey isn't only about acquiring intellectual knowledge; it's also about undergoing a profound shift in consciousness. As seekers pull back the veils of illusion, they come to realize the inherent oneness of all existence — that Brahman isn't merely a concept to grasp but a living reality to be experienced.

In the end, Brahma Vidya isn't just a path; it's a sacred pilgrimage of the soul — a quest for truth, wisdom, and spiritual liberation. Through the pursuit of Brahma Vidya, seekers aim to transcend the confines of the egoic mind and merge with the boundless expanse of Brahman. It's a journey of boundless insight, profound transformation, and ultimate liberation — a journey that leads to the realization of our true nature and the attainment of spiritual freedom.

The main avenues of exploration

What are the main avenues of exploration on this sacred path? Let's discover some of its mystique below.

Key aspects of Brahma Vidya include:

A. **Scriptural Study:** Practitioners engage in the systematic study of Vedantic scriptures such as the Upanishads, Bhagavad Gita, Brahma Sutras, and other relevant texts. These scriptures contain profound teachings on the nature of Brahman, the self, and the universe, serving as a guide for seekers on the path of Brahma Vidya.

B. **Philosophical Inquiry:** Brahma Vidya involves rigorous philosophical inquiry into the fundamental questions of existence, such as the nature of reality, the cause of the universe, and the essence of the self. Practitioners engage in deep contemplation and critical analysis to unravel the ultimate truth underlying all phenomena.

C. **Self-Inquiry:** Central to Brahma Vidya is the practice of self-inquiry (atma vichara), wherein practitioners introspectively investigate the nature of the self (Atman). By questioning the nature of identity, consciousness, and existence, seekers aim to penetrate the layers of illusion and realize the ultimate reality of Brahman within themselves.

D. **Meditation:** Meditation plays a crucial role in Brahma Vidya, facilitating the direct experience of Brahman beyond the realm of words and concepts. Practitioners engage in various meditation techniques designed to quiet the mind, deepen concentration, and open themselves to the intuitive apprehension of the absolute truth.

E. **Guru Discipleship:** Guidance from a qualified spiritual teacher is essential in Brahma Vidya. The guru imparts the teachings of Vedanta, provides clarification on complex philosophical concepts, and guides the disciple on the inner journey toward self-realization. The relationship between the guru and disciple is characterized by trust, devotion, and surrender.

Brahma Vidya represents an advanced path within Vedanta philosophy, suited for sincere seekers who are intellectually inclined and deeply committed to the pursuit of spiritual truth. Through dedicated study, contemplation, and meditation, the aim is to transcend the limitations of the conditioned mind and directly experience the eternal reality of Brahman, thereby attaining liberation (moksha) from the cycle of birth and death.

The source texts

1. Some of the key Upanishads that expound Brahman Vidya include:

Chandogya Upanishad: This Upanishad explores the nature of Brahman through various philosophical discussions and allegories. It contains the famous dialogue between Uddalaka Aruni and his son Svetaketu, where the father imparts the knowledge of Brahman to his son.

> *"That which is the subtle essence, in it all that exists has its self. It is the True. It is the Self, and thou, O Svetaketu, art it."* (Chandogya Upanishad 6.8.7)

This quote emphasizes the concept of the underlying unity of all existence. It suggests that there is a subtle essence, referred to as the True or the Self, which pervades everything in the universe. This essence is identified as Brahman, the ultimate reality. The Upanishad tells Svetaketu that he himself is this essence, implying that each individual is fundamentally connected to the cosmic whole.

Brihadaranyaka Upanishad: One of the oldest and largest Upanishads, the Brihadaranyaka Upanishad delves into profound metaphysical and philosophical teachings, including the nature of Brahman and the relationship between the individual self (Atman) and the universal Self (Brahman).

> *"Brahman is the immortal, supreme, eternal Soul. Brahman is the origin and support of everything. Brahman is the essence of the universe. Knowing this Brahman, one attains the highest."* (Brihadaranyaka Upanishad 4.4.25)

This passage reveals the nature of Brahman as the immortal, supreme, and eternal Soul. It asserts that Brahman is the source and sustainer of all existence, the essence of the universe itself. The text suggests that by knowing Brahman, one attains the highest state of realization, implying liberation (moksha) from the cycle of birth and death.

Mandukya Upanishad: This Upanishad focuses specifically on the nature of consciousness and the significance of the syllable "Om" as a symbol of Brahman. It describes four states of consciousness and elucidates the concept of Turiya, the transcendent state beyond waking, dreaming, and deep sleep.

> *"This syllable Om is the highest. Whoever knows this syllable obtains all that he desires. This is the best support; this is the highest support. Whoever knows this support is adored in the world of Brahman."* (Mandukya Upanishad 1.8)

This passage highlights the significance of the syllable "Om" as representing Brahman, the ultimate reality. It suggests that those who understand the profound meaning of Om can attain all their desires and achieve the highest fulfillment. Om is portrayed as the best support and the highest reality, and those who comprehend its essence are revered in the world of Brahma, signifying their spiritual attainment.

Taittiriya Upanishad: This Upanishad contains the famous Taittiriya-Vidya, a series of teachings on the nature of Brahman and the means to realize it. It emphasizes the importance of meditation, self-discipline, and moral conduct in attaining Brahman knowledge.

> *"Brahman is Bliss. On account of bliss, verily, these beings live, move, and have their being. Bliss is Brahman."* (Taittiriya Upanishad 3.6.1)

The quote equates Brahman with Bliss, emphasizing the inherent nature of Brahman as pure joy or happiness. It suggests that beings derive their life, movement, and existence from this Bliss, indicating that Brahman is the source of all happiness and fulfillment. The text highlights the divine nature of Brahman and its role in sustaining the world.

Katha Upanishad: The Katha Upanishad contains the profound dialogue between the young seeker Nachiketa and the god of death, Yama. In this dialogue, the nature of the eternal Self (Atman) and the path to realizing it are described.

> *"The Self-existent pierced the senses to turn outward. Thus we look to the outside and see not within. A sage withdrew his senses from*

> *the world of change and, seeking immortality, looked within and beheld the deathless Self."*
> (Katha Upanishad 2.1.1)

Here, we read of the contrast between the outward-directed nature of human senses with the inward journey of self-realization. It describes how the Self-existent (Brahman) created the senses to perceive the external world but suggests that true wisdom lies in turning inward to realize the immortal Self. The sage mentioned in the passage withdraws their senses from the changing world and discovers the eternal Self within, symbolizing the path to spiritual enlightenment and liberation.

These Upanishads, along with others like the Isha, Aitareya, and Mundaka Upanishads, form the foundational texts for Brahman Vidya practices. They contain timeless wisdom and spiritual insights that guide seekers on the path of self-realization and liberation through the knowledge of Brahman.

2. In the Bhagavad Gita, we also find nuggets of wisdom that align with the spiritual paths of Vedanta:

In one verse, we can read the following:

> *"Of the Vedas I am the Sama Veda; of the demigods I am Indra, the king of heaven; of the senses I am the mind, and in living beings I am the living force [consciousness]."*
> (Bhagavad Gita 10.22)

In this verse, Lord Krishna, who is speaking to Arjuna, reveals his divine manifestations in various aspects of creation. He says that among the Vedas , he is the Sama

Veda, emphasizing the importance and significance of this particular Veda.

Additionally, he identifies himself with Indra, the king of the demigods, indicating his supremacy among celestial beings. Furthermore, Krishna declares that among the senses, he manifests as the mind, highlighting the central role of the mind in human perception and cognition.

Finally, he asserts that in all living beings, he is the vital life force, which is consciousness itself. This verse underscores the divine presence in every aspect of creation, emphasizing the unity and interconnectedness of all beings with the divine.

Here's another relevant quote from the Bhagavad Gita:

> *"One who sees inaction in action, and action in inaction, is intelligent among men, and he is in the transcendental position, although engaged in all sorts of activities."* (Bhagavad Gita 4.18)

Lord Krishna explains the concept of detached action to Arjuna. He states that one who sees inaction in action understands the true nature of reality and is considered intelligent among men. This means that a spiritually wise person understands that their true self (Atman) is unaffected by the actions performed by the body and mind. From this perspective, even though engaged in various activities, the wise person remains unaffected and maintains inner tranquility.

This verse highlights the importance of performing one's duties without attachment to the results, focusing instead on selfless service and inner equanimity. It teaches that true spiritual intelligence lies in transcending the ego and recognizing the underlying unity of all existence.

3. The Brahmasutras also hold key insights into the path of enlightenment for the seeker:

> *"Now, therefore, the inquiry into Brahman."* (Brahmasutra 1.1.1)

This opening sutra sets the stage for the entire text, indicating that the primary subject of inquiry is Brahman, the ultimate reality. It underscores the importance of delving into the nature of Brahman as the central theme of Vedanta philosophy.

> *"And there is absence of this, on account of the declaration of difference."* (Brahmasutra 1.3.33)

This sutra addresses the question of whether the individual self (jiva) and Brahman are identical or different. It emphasizes the distinction between the individual self and Brahman, asserting that they are not identical but rather qualitatively different.

> *"Sound is Brahman, on account of the declaration of attributes."* (Brahmasutra 2.1.14)

This sutra discusses the nature of Brahman as it relates to sound (Om). It affirms that sound is a manifestation of Brahman and not merely a mundane phenomenon. This highlights the omnipresence and transcendental nature of Brahman.

These quotes from the Brahmasutras provide a glimpse into the depth and complexity of Vedanta philosophy, offering insights into the nature of Brahman, the relationship between the individual self and the ultimate reality, and the fundamental principles of creation.

Key concepts

As I mentioned earlier, Brahma Vidya isn't just an ordinary philosophical pursuit. It's a cosmic odyssey into the very essence of existence itself. Imagine diving headfirst into the deep end of the universe, armed with nothing but profound questions and an insatiable thirst for truth.

This sacred inquiry doesn't shy away from the big questions that have perplexed humanity since time immemorial. It's a daring quest to unravel the mysteries that lie at the heart of creation. What is the nature of reality? What is the essence of consciousness? These are the profound questions that Brahma Vidya seeks to answer, plunging into the depths of the unknown with fearless curiosity.

At its core, Brahma Vidya is a journey of discovery — a relentless pursuit of the ultimate truth that underlies all existence. It's a quest that transcends the boundaries of ordinary perception, inviting seekers to peer beyond the veil of illusion and glimpse the infinite expanse of reality.

So, buckle up and prepare for an exhilarating ride as we embark on this cosmic adventure through the realms of Brahma Vidya. It's a journey that promises to challenge your perceptions, expand your consciousness, and ultimately lead you to the profound realization of your true nature.

Here are some examples of those questions:

What is the Nature of Reality?

Brahma Vidya is a profound exploration into the essence of reality, transcending appearances to seek the ultimate truth. Practitioners contemplate Brahman, the boundless reality underlying all existence. Through meditation and introspection, they strive to connect with Brahman and uncover the unity within diversity. Despite challenges, this journey offers spiritual insights and transformative growth, leading seekers beyond illusion to pure consciousness and self-discovery.

> The Chandogya Upanishad (6.8.7) declares:
> *"That which is the subtle essence, in it all that exists has its self. It is the True. It is the Self, and thou, O Svetaketu, art it."*

This quote emphasizes the interconnectedness of all existence within the subtle essence of Brahman, inviting seekers to contemplate the unity underlying the diversity of the world.

What is the Cause of the Universe?

Brahma Vidya is a spiritual quest to uncover the fundamental cause of the universe's existence. Practitioners embark on a journey of deep contemplation, pondering whether the cosmos arose from a divine origin or emerged from another principle altogether. This ancient pursuit invites seekers to explore the mysteries of creation and the nature of ultimate reality. The Brihadaranyaka Upanishad (4.4.25) asserts:

> *"Brahman is the immortal, supreme, eternal Soul. Brahman is the origin and support of everything. Brahman is the essence of the*

> *universe. Knowing this Brahman, one attains the highest."*

This quote highlights Brahman as the ultimate cause and support of the universe, inviting one to contemplate the cosmic principle underlying all phenomena.

What is the Essence of the Self?

In addition, Brahma Vidya entails a profound exploration of the self, delving into the essence of individual identity beyond the confines of the body and mind. Practitioners strive to grasp the true nature of the self, transcending conventional notions to uncover the deeper layers of consciousness and existential reality. The Mandukya Upanishad (1.8) proclaims:

> *"This syllable Om is the highest. Whoever knows this syllable obtains all that he desires. This is the best support; this is the highest support. Whoever knows this support is adored in the world of Brahma."*

This quote emphasizes the significance of the sacred syllable Om as a symbol of the ultimate reality, inviting seekers to contemplate their essential identity with Brahman.

What is the Unity of Existence?

Vedanta emphasizes the underlying unity of all existence, suggesting that there is a fundamental essence (Brahman) that pervades everything in the universe. This concept challenges the notion of separateness and highlights the interconnectedness of all beings. The Katha Upanishad (2.1.1) declares:

> *"The Self-existent pierced the senses to turn outward. Thus we look to the outside and see not within. A sage withdrew his senses from the world of change and, seeking immortality, looked within and beheld the deathless Self."*

This quote elucidates the journey from outward perception to inner realization of the eternal Self. It suggests that by transcending the outward-oriented senses and turning inward, one can perceive the underlying unity beyond the diversity of the world. This inward journey leads to the direct experience of the deathless Self, which transcends the cycle of birth and death, revealing the eternal unity of existence.

What is the Path to Liberation?

The insights within sacred texts allude to the path to liberation (moksha) or spiritual awakening, which involves turning inward and realizing the true nature of the Self. This journey requires transcending the outward-directed senses and recognizing the eternal Self within.

> *"That which is the subtle essence, in it all that exists has its self. It is the True. It is the Self, and thou, O Svetaketu, art it."* (Chandogya Upanishad (6.8.7)

This quote emphasizes the subtle essence (Brahman) as the true Self that pervades all existence. The passage suggests that individuals, like Svetaketu, are inherently one with this essence. By recognizing and realizing the true nature of the Self as Brahman, individuals can attain liberation from the cycle of birth and death. This realization involves turning inward to perceive the eternal essence within oneself and in all beings.

What is the Importance of Spiritual Knowledge?

Knowledge (Jnana) plays a crucial role in attaining liberation. Understanding the nature of Brahman and the Self is depicted as the key to ultimate fulfillment and freedom from the cycle of birth and death.

In the Bhagavad Gita, the seeker learns of the importance of spiritual knowledge in attaining liberation. In Chapter 4, Verse 38, Lord Krishna instructs Arjuna,

> *"In this world, there is nothing so purifying as spiritual knowledge. He who has attained purity of heart through prolonged practice of Karma Yoga automatically sees the light of truth in the self."*

This verse highlights the purifying and enlightening nature of spiritual knowledge, which leads to self-realization and liberation from the cycle of birth and death.

What is the Role of Consciousness?

Consciousness is implied as the essence of Brahman and the source of all existence. By turning inward and realizing the deathless Self, one taps into the realm of pure consciousness beyond the limitations of the material world.

The Mandukya Upanishad is dedicated to the exploration of consciousness and the nature of reality. In its verses, it describes four states of consciousness: waking (jagrat), dreaming (svapna), deep sleep (sushupti), and the fourth state (turiya) beyond the three. It emphasizes that the fourth state, characterized by pure consciousness, is the true essence of the Self. For example, in Verse 7, it states,

> *"This is the Lord of all, the knower of all, the inner controller, the source of all, the origin and dissolution of all beings."*

This verse highlights the role of pure consciousness as the source and essence of all existence.

While the concepts themselves may seem abstract and challenging to comprehend intellectually, they may resonate deeply on an intuitive and experiential level. Through repeated contemplation, study, and spiritual practice, individuals can fully grasp and internalize these profound teachings.

Brahma Vidya is a profound philosophical inquiry into the deepest mysteries of existence, inviting practitioners to engage in contemplation, reflection, and introspection to unravel the ultimate truth underlying all phenomena. Through this rigorous inquiry, seekers endeavor to attain direct insight into the nature of reality, the cause of the universe, and the essence of the self, ultimately leading to spiritual awakening and liberation.

Overall, the journey of understanding these teachings is a deeply personal and transformative process, often unfolding gradually over time.

What does a direct experience of the boundless expanse of Brahman feel like?

The direct experience of the boundless expanse of Brahman is often likened to diving into an ocean of infinite consciousness. Imagine being submerged in a sea of profound tranquility, where every ripple of thought dissolves into a vast expanse of pure awareness. It's a feeling of boundless expansion, where the limitations of the individual self melt away, and you become one with the cosmic pulse of existence.

This experience is often accompanied by a sense of profound peace and bliss, akin to basking in the warm glow of a thousand suns. It's a state of absolute stillness, where the incessant chatter of the mind fades into silence, and you are enveloped in a timeless embrace of pure being.

Moreover, the direct experience of Brahman is said to bring about a profound sense of interconnectedness with all of creation. It's like realizing that you are not just a drop in the ocean, but the ocean itself, inseparable from the fabric of existence. In this state of heightened awareness, the boundaries between self and other dissolve, and you perceive the underlying unity that binds all beings together.

Overall, the direct experience of Brahman is described as a transcendent journey beyond the confines of ordinary perception — a glimpse into the infinite expanse of consciousness that lies at the heart of reality. It's a profound awakening that leaves a lasting impression, forever transforming the way you perceive yourself and the world around you.

Some common aspects include:

- **Unity and Oneness:** Those who have had glimpses of Brahman often describe a sense of unity and oneness that surpasses the distinctions of the everyday world. The boundaries between the observer and the observed, subject and object, seem to dissolve.

- **Ineffable Bliss:** The experience is often associated with an overwhelming sense of bliss or joy, which is said to surpass any mundane happiness. This bliss is not dependent on external circumstances but is intrinsic to the nature of Brahman.
- **Transcendence of Time and Space**: The boundless expanse of Brahman is beyond the limitations of time and space. Experiencing Brahman involves a sense of transcendence, where the ordinary constraints of past, present, and future dissolve.
- **Eternal Presence:** Brahman is often described as an eternal, ever-present reality. Those who have direct experiences of Brahman may feel a connection to an enduring and timeless dimension of existence.
- **Beyond Individual Identity**: The experience of Brahman involves a transcendence of individual identity. The egoic sense of self dissolves, and there is a merging or identification with a universal, formless consciousness.
- **Profound Stillness:** Brahman is often associated with a state of profound stillness. In this experience, the fluctuations of the mind and the restless nature of ordinary consciousness come to a profound rest.

What does that inner stillness and awareness feel like?

The experience of inner stillness and awareness cultivated through practices like meditation can indeed vary greatly from person to person and even from session to session. It's a deeply personal journey, influenced by individual temperament, state of mind, and level of practice.

For some, meditation may bring a profound sense of calm and tranquility right from the start. They may effortlessly slip into a state of deep relaxation, where the mind becomes quiet, and the body feels light and at ease. In this state, there's a heightened awareness of the present moment, as if time slows down, and one becomes fully immersed in the here and now.

However, for others, the path to inner stillness may be more challenging. The mind may be restless, filled with thoughts, worries, and distractions that seem to pull attention in all directions. In such cases, the practice of meditation becomes a gentle process of returning to the present moment again and again, patiently guiding the mind back to a state of calmness and clarity.

Despite these variations, there are common threads that many practitioners experience during meditation. These include a sense of deep relaxation and physical ease, a heightened awareness of bodily sensations and breath, and moments of mental clarity and insight. Over time and with consistent practice, these experiences tend to deepen, leading to a more profound sense of inner peace, presence, and connection.

Here are some common aspects that practitioners often describe:

- **Deep Peace:** Many people report a profound sense of peace and tranquility during meditation. It's as if the mind becomes calm and serene, free from the usual mental chatter and turmoil.

- **Clarity of Mind:** As the mental noise diminishes, there can be a heightened clarity of thought and perception. It may feel like the fog of confusion is lifting, allowing one to see things more clearly, both internally and externally.
- **Expanded Awareness**: Some describe a feeling of expansiveness or openness during meditation as if their consciousness is expanding beyond the confines of their physical body. This expanded awareness can lead to a sense of connection with something greater than oneself.
- **Heightened Sensitivity:** With the mind quieted, practitioners may become more attuned to subtle sensations within the body and subtle aspects of their environment. This heightened sensitivity can deepen the experience of being present in the moment.
- **Profound Stillness:** At deeper levels of meditation, there can be moments of profound stillness where the sense of individual identity dissolves, and one feels connected to a vast, boundless presence. This state is often described as pure being or pure awareness.

It's important to note that these descriptions are subjective and can vary based on individual experiences and stages of practice. Some may experience these states more readily, while others may require more time and effort to access them. Additionally, the depth of meditation experiences can deepen over time with consistent practice.

Brahma Vidya holds the power to transform our lives in profound and inspiring ways, opening ourselves to boundless possibilities. Through deep contemplation, meditation, and connection with the divine, we expand our consciousness, nurture spiritual growth, and find inner peace and fulfillment. With each step, we gain clarity, wisdom, and a deeper sense of purpose, aligning ourselves with the universal principles that govern existence.

Brahma Vidya invites us to awaken to the inherent beauty and divinity within ourselves and the world around us, empowering us to live with intention, authenticity, and grace. By embracing the teachings of Brahma Vidya, we discover the infinite potential of our very being and shine as beacons of love, light, and truth in the world.

Moving forward to our next chapter, we will delve into the intricate practices of Tantra, offering further insights into harnessing spiritual energies and expanding our consciousness. Intrigued? Let's go, then!

Once again we meet

During one of my trips to Egypt, I stumbled upon a chance meeting that would forever alter my relationship with this region. While exploring the ancient Temple of Horus, I encountered a mysterious figure — a local healer known for his deep connection to the sacred energies that permeated the land.

As we exchanged greetings, I felt a sense of familiarity and intrigue wash over me. Sensing my curiosity, the healer's eyes sparkled with recognition, as if he had been expecting me. With a warm smile, he greeted me by name. *How could he know that?*

On delving deeper into the heart of the temple, the healer shared stories of our shared past — a past life in ancient Egypt, where we had been fellow students, bound by a sacred oath to protect and preserve the ancient wisdom of the land.

In the inner alcoves and chambers, the healer pointed out ancient symbols etched into the stone walls, their meanings intertwined with the threads of our shared history. With each revelation, I found myself drawing parallels between the sacred symbols of Egypt and those of the Vedic tradition — symbols that spoke of universal truths and timeless wisdom, passed down through the ages by ancient sages and seers. I did my best to explain similar symbolism and rituals from the Vedas to the healer as well. He was also very curious, asked a lot of questions and seemed very grateful to learn of these similarities. With every passing moment, what struck me the most was the striking resemblance between the symbols and rituals of ancient Egypt and those of the Vedic tradition — a connection that transcended time and space, weaving a tapestry of unity between distant cultures.

As the day waned and the sun dipped below the horizon, we emerged from the depths of the temple, our hearts heavy with the weight of the knowledge we had gained. Yet, amidst the fading light, I felt a newfound sense of purpose stirring within me — a deeper understanding of the interconnectedness of all things, and the profound wisdom that lay hidden within the symbols and rituals of ancient cultures. I know one day I have to write about it. I still feel I will, maybe this is the year to do it!

Bidding farewell to the healer and the Temple of Horus, I carried with me not only memories of our chance encounter but also a profound gratitude for the glimpse into the mysteries of the past that had been revealed to me. I hope my path crosses with that healer again. It will — if it's meant to be!

"Tantra is a way of life, a way of thinking and feeling, that leads to the realization of the divine within oneself and in the universe."

– Alain Daniélou

CHAPTER 16

TANTRA: HARNESSING ENERGIES FOR SPIRITUAL AWAKENING

You are probably familiar with the term 'Tantra' but aren't quite sure what it really means. Tantra may seem mysterious, enigmatic, or even exotic, due to the word having been highly misconstrued in the West.

The misconception stems largely from the focus on Tantra's more esoteric aspects, particularly those related to sexuality. While Tantra does incorporate practices that acknowledge and honor the body's energies, including sexual energy, these practices are intended to be approached with reverence, discipline, and spiritual intention. They are not meant to be interpreted as mere indulgence in carnal pleasures.

What is Tantra?

Tantra is actually a fascinating and intricate path within Vedanta dedicated to unlocking spiritual potential. Imagine it as a kaleidoscope of practices, blending **rituals, mantra recitations**, **visualizations**, **meditation,** and **breath control technique**s to awaken the lifeforce of spiritual energy (kundalini) and attain enlightenment.

The word itself is derived from Sanskrit, with "Tantra" (तन्त्र) derived from the root word "tan," which means "to weave" or "to expand." Thus, Tantra can be interpreted to mean "a system" or "a doctrine" that is woven together. This weaving metaphor suggests the interconnectedness and integration of various aspects of life and spiritual practices within Tantra.

In the context of spiritual traditions, Tantra refers to a diverse range of esoteric practices, philosophies, and rituals aimed at spiritual growth, liberation, and the realization of divine consciousness. It encompasses a holistic approach to spirituality that integrates physical, mental, emotional, and spiritual dimensions of existence.

At its core, Tantra perceives the universe as a canvas painted with divine energy, where every brushstroke reveals the unity between the individual self (Atman) and the cosmic consciousness (Brahman). It's about transcending dualities and limitations, tapping into the boundless reservoir of divine essence within.

In essence, Tantra is a profound spiritual path that goes beyond superficial interpretations and sensationalized depictions. It offers a holistic approach to spiritual growth and self-realization, inviting practitioners to delve deeply into the mysteries of existence.

What sets Tantra apart is its recognition of the senses, emotions, and even the physical body as gateways to spiritual realization. It's like inviting every aspect of your being to dance to the rhythm of cosmic harmony, infusing mundane existence with divine purpose.

Central to Tantra is the worship of deities, not as external beings, but as embodiments of divine qualities. Through intricate rituals and visualizations, practitioners invoke the presence of these deities to cultivate virtues like love, compassion, and wisdom within themselves. It's akin to sculpting the soul with divine clay, shaping it into a masterpiece of spiritual evolution. Tantra also utilizes sacred symbols known as yantras and mystical syllables called mantras to focus the mind and channel spiritual energy.

Breath control techniques (pranayama) are integral to Tantra practices, as they help regulate the flow of vital energy (prana) in the body and awaken the dormant spiritual energy located at the base of the spine (kundalini). By mastering the breath and directing it through specific energy channels (nadis), practitioners can purify the body and mind, awaken higher states of consciousness, and experience spiritual ecstasy.

And then there's meditation, the heart of Tantra, where the mind becomes a canvas for painting cosmic visions. Practitioners dive deep into the ocean of consciousness, navigating through subtle energy centers (chakras) within the body, like explorers charting the course to spiritual union. Through meditation practices, practitioners learn to quiet the mind, transcend ordinary perception, and merge with the universal consciousness.

In essence, Tantra is a sacred alchemy, transforming mundane existence into a tapestry of divine revelation. It's about embracing the totality of human experience and recognizing the divinity pulsating within every atom of creation. Through Tantra, seekers embark on a transformative odyssey, unraveling the mysteries of existence and attaining the ultimate enlightenment.

The practices of Tantra are aimed at harmonizing the dualistic aspects of existence, such as masculine and feminine energies, and transcending the limitations of the material world to realize the divine essence within. Tantra also recognizes the power of the senses, emotions, and physical body as vehicles for spiritual realization and emphasizes the integration of all aspects of human experience into the spiritual path.

What is tantric sexuality?

One of the most well-known books about tantric sexuality is the "Mahanirvana Tantra," translated in 1913 by Sir John Woodroffe under the pseudonym 'Arthur Avalon'. He was one of the few Indologists to gain direct access to this obscure and secretive branch of Hinduism, helping to demystify Tantra and present it as a profound spiritual path for personal transformation and self-realization.

Here are some of the insights and techniques related to sacred sexuality mentioned in his book:

Sublimation of Sexual Energy: The practice of sublimating sexual energy is about learning to redirect the powerful force of sexual desire towards spiritual pursuits. By channeling this energy upwards through the subtle energy channels (nadis) of the body, practitioners aim to awaken higher states of consciousness and activate the dormant kundalini energy.

Tantric Rituals for Union: The "Mahanirvana Tantra" outlines specific tantric rituals and ceremonies designed to facilitate spiritual union between partners. These rituals often involve symbolic gestures, mantras, and visualization techniques aimed at harmonizing the energies of both individuals and transcending the physical realm to connect with the divine.

Maithuna Practice: Maithuna, or sacred sexual union, is a tantric practice for deepening intimacy and spiritual connection between partners. It emphasizes mutual respect, love, and reverence for the divine presence within each other. Through mindful and heart-centered lovemaking, practitioners seek to merge their energies and experience a profound sense of oneness with the divine.

Breathwork and Sensory Awareness: The "Mahanirvana Tantra" explores the use of breathwork (pranayama) and heightened sensory awareness during sacred sexual practices. By synchronizing breath and movement, practitioners aim to deepen their connection with the present moment, heighten sensations, and amplify the flow of energy throughout the body.

Transcendence of Ego and Dualism: Through the practice of sacred sexuality, practitioners can transcend the limitations of the ego and dualistic thinking. By experiencing the dissolution of boundaries between self and other during lovemaking, individuals can attain glimpses of unity consciousness and the interconnectedness of all existence.

While the Mahanirvana Tantra contains various chapters dedicated to different aspects of Tantra, including sexual practices known as Maithuna or Tantra Yoga, these chapters typically discuss the union of masculine and feminine energies (Shiva and Shakti) as a means of spiritual awakening and enlightenment. We also need to approach the text with an understanding of its cultural and historical context and not solely as a manual for sexual techniques. Indeed, the proportion of the text that explicitly discusses sexual energy varies, as the Mahanirvana Tantra covers a wide range of topics related to Tantra. The sexual aspect is a significant part, but it is not the sole focus of the text.

Tantra sources

The practices of Tantra draw from a variety of ancient scriptures, Tantric texts, and spiritual traditions. Some of the key sources for Tantra include:

1. **Tantras:** Tantras are a genre of Hindu scriptures that form the foundation of Tantric philosophy and practices. They include texts such as the Mahanirvana Tantra we already mentioned, as well as the Kularnava Tantra, Shiva Samhita, and the Vijnana Bhairava Tantra, among others. These texts contain teachings on rituals, meditation techniques, deity worship, and spiritual philosophy.

    a. The **Mahanirvana Tantra** expounds upon the worship of Shakti, the divine feminine energy, and provides guidance on mantra chanting, yantra usage, and the significance of inner purification. In this classical Tantric text, Shakti is often personified as the Goddess and represents the creative and transformative energy of the universe. The text denotes Shakti as a central aspect of Tantra, emphasizing her role in spiritual awakening and liberation, and providing detailed instructions on various forms of Shakti worship.

    b. The **Kularnava Tantra** is revered for its comprehensive insights into Tantric spirituality. It provides detailed instructions on deity worship, offering a wealth of rituals and ceremonies for practitioners. Through mantra recitation and meditation, it guides seekers in awakening higher states of consciousness, ultimately leading to liberation from worldly limitations and union with the divine. The text's teachings emphasize the integration of spiritual practices into daily life, fostering profound inner transformation and a deeper understanding of the universal truths of existence.

c. The **Shiva Samhita** is a classical text that provides in-depth guidance on Hatha Yoga practices. It covers various aspects such as asanas, pranayama, and mudras, all aimed at cleansing the body and mind while awakening latent spiritual energies. Through these practices, practitioners seek to achieve physical vitality, mental clarity, and spiritual awakening. The text emphasizes the importance of discipline, proper technique, and inner focus to harness the transformative power of Hatha Yoga effectively.

d. The **Vijnana Bhairava Tantra** is a profound text featuring a dialogue between Lord Shiva and Goddess Bhairavi. It delves into diverse meditation techniques and states of consciousness aimed at achieving spiritual realization. Through this dialogue, practitioners gain insights into methods for accessing deeper levels of awareness and transcending the limitations of the ego. The text serves as a guide for seekers on the path of self-discovery and enlightenment, offering timeless wisdom to navigate the complexities of the human experience.

These Tantric scriptures not only provide practical guidance on rituals and meditation but also expound profound spiritual philosophy. They emphasize the immanence of the divine within all beings, the power of mantra and yantra in spiritual practice, and the recognition of the ultimate reality beyond dualities. Through the study and application of Tantric teachings, practitioners seek to awaken their innate divine potential and attain union with the Supreme.

2. **Agamas:** The Agamas are a rich collection of scriptures, offering comprehensive guidelines on temple construction, rituals, and worship practices. These texts are dedicated to various deities and serve as manuals for conducting elaborate ceremonies aimed at invoking divine energies and facilitating spiritual evolution. Some examples are mentioned below.

    a. The **Kamika Agama** provides detailed instructions on temple architecture, consecration rituals, and daily worship procedures. It emphasizes the importance of precise measurements, auspicious alignments, and sacred symbols in constructing a temple conducive to spiritual practices. The text also elucidates on various rituals for deity worship, offering insights into the significance of offerings, mantras, and mudras in invoking divine presence

    b. The **Vaikhanasa Agama** is a sacred text devoted to the worship of Lord Vishnu within the Vaishnavite tradition. It provides detailed guidelines for rituals conducted in Vaishnavite temples, covering aspects such as deity installation, daily worship practices, and festive celebrations. Emphasizing devotion and discipline, the text underscores the importance of sincerity and precision in worship to foster a deeper connection with the divine.

The Agamas also include texts dedicated to other deities such as Shiva, Devi (the divine feminine), and Ganapati (Lord Ganesha), each providing unique insights into their worship practices. For example, the Karana Agama elaborates on the rituals associated with Shiva worship, including the performance of abhisheka (ritual bathing) and the chanting of sacred hymns like the Shiva Sahasranama.

These scriptures serve as invaluable guides for priests, temple architects, and devout practitioners, offering a comprehensive framework for engaging in devotional activities and spiritual rituals. By adhering to the teachings of the Agamas with faith and sincerity, devotees seek to deepen their connection with the divine and attain spiritual fulfillment.

3. Upanishads: While the Upanishads primarily focus on philosophical inquiry and spiritual wisdom, some Tantric interpretations draw inspiration from certain Upanishadic teachings. The Chandogya Upanishad, Brihadaranyaka Upanishad, and Maitrayaniya Upanishad are among those Upanishads that have influenced Tantra philosophy.

a. The **Chandogya Upanishad** contains profound discussions on the nature of Brahman and the self, presenting teachings that emphasize the unity of existence and the underlying interconnectedness of all beings. This concept resonates deeply with Tantra philosophy, which often emphasizes the recognition of divinity in all aspects of creation. Tantric practitioners may draw inspiration from the Chandogya Upanishad's teachings on the ultimate reality of Brahman to cultivate a sense of unity and reverence for all manifestations of the divine.

b. The **Brihadaranyaka Upanishad** also expounds upon the nature of Brahman as the supreme, eternal soul, and the essence of the universe. It discusses the role of consciousness in the creation and sustenance of the cosmos, presenting profound insights into the interconnected nature of reality. Tantric practitioners may find resonance with these teachings, incorporating them into their contemplative practices to deepen their understanding of the divine nature of existence.

c. The **Maitrayaniya Upanishad** delves into the concept of the self (Atman) and its relationship with Brahman, presenting teachings on the inherent divinity and immortality of the soul. Tantric interpretations of these teachings may emphasize the realization of the divine essence within oneself and the cultivation of spiritual practices aimed at realizing one's true nature as Brahman.

In Tantric philosophy, these Upanishadic teachings serve as sources of inspiration and guidance for spiritual seekers on the path of self-realization. By drawing from the profound wisdom contained within them, practitioners of Tantra seek to deepen their spiritual understanding and experience profound states of consciousness and union with the divine.

4. **Puranas:** The Puranas are a genre of ancient Hindu texts that contain myths, legends, and narratives about gods, goddesses, and spiritual concepts. While not explicitly Tantric in nature, some Puranic texts contain elements of Tantra and provide stories and descriptions of Tantric practices.

    a. For example, the **Shiva Purana** and the **Devi Bhagavata Purana** are among the Puranic texts that prominently feature narratives related to Tantric themes and practices. These texts depict stories of Lord Shiva and the Goddess Devi engaging in mystical rituals, meditation, and worship, often incorporating esoteric Tantric elements such as mantra recitation, yantra worship, and Kundalini awakening.

    b. In the Shiva Purana, tales of Shiva's divine manifestations and cosmic dances symbolize profound Tantric concepts such as the interplay of Shiva and Shakti, the union of masculine and feminine energies, and the transformative power of spiritual practices.

    c. Similarly, the Devi Bhagavata Purana elaborates on the worship of the Divine Mother (Devi) in her various forms, emphasizing the esoteric rituals and meditative techniques employed by devotees to attain spiritual liberation.

While the Puranas primarily serve as repositories of mythological lore and devotional narratives, they also offer glimpses into the world of Tantra through their descriptions of sacred rituals, hymns, and philosophical discussions. These Tantric elements provide practitioners and scholars with valuable insights into the historical development and cultural context of Tantric traditions within Hinduism.

Overall, while the Puranas are not exclusively Tantric texts, their inclusion of Tantric themes and practices underscores the interconnectedness of various strands of Hindu spirituality and highlights the diverse array of spiritual paths available to seekers throughout history.

5. Smriti Texts: Smriti texts are an extensive body of Hindu scriptures that complement the Vedas and are considered authoritative sources for guiding religious and social conduct. Among the various Smriti texts, the Dharmashastras and the Grihya Sutras stand out as significant sources of guidance on religious duties, social ethics, and ritualistic practices.

While these texts are not exclusively dedicated to Tantra, they occasionally contain references to Tantric principles, practices, and beliefs, albeit in a more implicit or indirect manner.

> a. For example, the **Manusmriti**, one of the most well-known Dharmashastras, offers guidelines on various aspects of Hindu life, including family structure, social hierarchy, and ethical conduct. While it primarily emphasizes Vedic rituals and social norms, some scholars have identified passages within the Manusmriti that reflect Tantric principles of ritual purity, meditation, and the worship of divine energies.

b. Similarly, the **Grihya Sutras**, which deal with domestic rituals and ceremonies, often include instructions for rites of passage, such as birth ceremonies, weddings, and funerals. While these texts predominantly focus on orthodox Vedic rituals, there are instances where Tantric elements, such as the use of mantras, yantras, and symbolic gestures (mudras), are incorporated into the performance of these rituals.

It's important to recognize that the inclusion of Tantric references in Smriti texts does not necessarily imply that these texts are exclusively Tantric or that they endorse Tantra as a distinct path. Instead, it reflects the syncretic nature of Hinduism, where diverse religious traditions and philosophical perspectives often intersect and influence one another. The occasional presence of Tantric elements in Smriti texts underscores the rich and multifaceted nature of Hindu religious literature and the complexity of its historical development.

6. Tantric Lineages: Tantra has been transmitted through oral tradition and lineage-based teachings within various Tantric schools and sects. Practitioners often receive teachings and instructions from Tantric gurus (teachers) who are part of specific Tantric lineages.

These sources, along with oral traditions and the guidance of Tantric gurus, form the basis of Tantric practices and teachings.

Tantra teachings

Let's take a look at some intriguing Tantra practices and teachings below.

"Om Namah Shivaya" - Kularnava Tantra

> This is a powerful mantra in Tantra dedicated to Lord Shiva. It translates to "I bow to Shiva," and it is considered a sacred chant for invoking the divine presence of Shiva, the supreme consciousness.

"Om Hrim Shrim Krim Parameshvari" - Mahanirvana Tantra

> This mantra is an invocation to the Divine Mother, the Supreme Goddess, who is addressed as Parameshvari. It is composed of seed syllables (bija mantras) representing the creative and transformative energies of the Goddess.

"Shivaya Vishnu Rupaya, Shiva Rupaya Vishnave" - Shiva Samhita

> This verse expresses the inseparable nature of Lord Shiva and Lord Vishnu, two principal deities in Hinduism. It signifies that Shiva is the embodiment of Vishnu and vice versa, highlighting their underlying unity beyond their apparent differences.

"Neti Neti" - Vijnana Bhairava Tantra

> This phrase means "not this, not this" and is used in the context of self-inquiry and meditation. It encourages the practitioner to negate all identifications with the transient aspects of existence, leading to the realization of the true Self beyond all conceptualizations.

"Satyam Jnanam Anantam Brahma" - Chandogya Upanishad 3.14.1

This statement from the Chandogya Upanishad declares that Brahman, the ultimate reality, is of the nature of truth, knowledge, and infinity. It emphasizes the transcendental nature of Brahman beyond the limitations of time, space, and causation.

"Tat Tvam Asi" - Brihadaranyaka Upanishad 1.4.10

This famous Mahavakya (great aphorism) means “That thou art” and signifies the identity of the individual self (Atman) with the ultimate reality (Brahman). It reveals the non-dual nature of existence, where the essence of the individual soul is identical to the supreme consciousness.

“Ayam Atma Brahma” - Maitrayaniya Upanishad 1.2.5

This phrase translates to “This Self is Brahman” and asserts the identity of the individual self with the ultimate reality, Brahman. It emphasizes the immanent presence of Brahman within the individual soul and underscores the inherent divinity of all beings.

These explanations provide insights into the profound spiritual truths conveyed in each quote, reflecting the core teachings of Tantra and Upanishads.

The Path of Tantra

The path of Tantra can be practiced through various methods aimed at harnessing energies for spiritual awakening. Each practice is rooted in a profound philosophical framework aimed at spiritual transformation and realization. Here are some key practices:

- **Mantra Meditation:** Mantras are sacred sounds, syllables, or phrases that are believed to have inherent spiritual power. The philosophy behind mantra meditation is based on the principle of sound vibrations (nada), which can resonate with the subtle energies of the practitioner's being and the cosmos, leading to purification, concentration, and spiritual upliftment. Chanting sacred mantras, such as "Om Namah Shivaya" or "Om Hrim Shrim Krim Parameshvari," with devotion and focused intention can awaken spiritual energies and lead to higher states of consciousness.
- **Yantra and Mandala Meditation:** Yantras and mandalas are geometric diagrams representing the divine forces or aspects of the cosmos. The philosophy behind their meditation lies in the concept of sacred geometry and the belief that visualizing and meditating upon these intricate patterns can awaken higher states of consciousness, align the practitioner with universal principles, and facilitate spiritual realization. Meditating on these geometric forms can help concentrate the mind and channel divine energies for spiritual growth and realization.
- **Kundalini Yoga:** Kundalini is the dormant spiritual energy believed to reside at the base of the spine, represented as a coiled serpent. Kundalini Yoga aims to awaken this latent energy through specific practices, such as asanas, pranayama, and meditation. The philosophy behind Kundalini Yoga is grounded

in the idea of harnessing the evolutionary potential within each individual, leading to spiritual enlightenment and union with the divine. Practicing specific yoga techniques, including asanas (physical postures), pranayama (breath control), and bandhas (energy locks), to awaken and channel the dormant spiritual energy.

- **Rituals and Puja:** Tantric rituals and puja involve the worship of deities, cosmic forces, or divine aspects through elaborate ceremonies, offerings, and invocations. The philosophy behind these practices is rooted in the belief that external rituals can serve as potent means of inner transformation, invoking divine blessings, purifying the mind, and fostering a deep sense of connection with the sacred. Engaging in these sacred rituals and worship ceremonies, such as fire ceremonies (homa) or deity worship (puja), can facilitate communion with divine energies and invoke their blessings for spiritual progress.
- **Meditative Visualization**: Tantric visualization practices involve mentally picturing and connecting with divine archetypes, energy centers (chakras), or symbolic representations of cosmic principles. The philosophy behind meditative visualization is based on the principle of "as above, so below," wherein the practitioner seeks to align their inner reality with the divine order of the universe, leading to expanded awareness and spiritual awakening.

- **Tantric Text Study:** Studying Tantric scriptures under the guidance of a qualified teacher (guru) is integral to understanding the principles of Tantra. The philosophy behind Tantric texts emphasizes the interplay of Shiva (consciousness) and Shakti (divine energy), the worship of the divine feminine (Shakti or Devi), and the recognition of the world as a manifestation of divine play (lila), all aimed at realizing the ultimate unity of existence. Through the study of sacred Tantric texts, such as the Tantras, Agamas, or Shakta scriptures, one can gain deeper insights into Tantra philosophy, symbolism, and practices.
- **Integration of Tantra Principles:** The philosophy behind integrating Tantric principles into daily life is based on the idea of seeing the divine in all aspects of existence and recognizing that every experience, whether mundane or spiritual, can be a path to enlightenment. Tantra emphasizes the holistic integration of body, mind, and spirit, leading to a harmonious and sacred way of living in the world.

The above practices aim to harmonize and balance the energies of the body, mind, and spirit, leading to self-realization and union with the divine. It's essential to approach Tantra with sincerity, reverence, and ethical integrity, following the guidance of authentic teachings and spiritual mentors.

Integration of Tantra Principles:

Sexual alchemy techniques in Tantra involve practices aimed at harnessing and transmuting sexual energy for spiritual growth and enlightenment. These techniques are often part of advanced Tantric practices and are approached with reverence, mindfulness, and a deep understanding of Tantra's spiritual principles. Here are some common elements and concepts related to sexual alchemy in Tantra:

- ♦ Kundalini as the Source of Spiritual Energy:
  - ◊ Kundalini, often depicted as a coiled serpent lying dormant at the base of the spine, is considered the primal cosmic energy within each individual. It is the source of spiritual energy and is closely associated with the awakening of higher consciousness.
  - ◊ "The Kundalini, which is the power of the omnipresent Divine, is dormant in the human body and is represented as a serpent coiled three and a half times round the svayambhu-linga (at the base of the spine) in the Muladhara-chakra." (The Serpent Power: The Secrets of Tantric and Shaktic Yoga by Arthur Avalon)
- ♦ Harnessing Sexual Energy for Spiritual Growth:
  - ◊ In Tantra, sexual energy is considered a potent force that can be harnessed and transmuted for spiritual evolution. Rather than dissipating this energy through ordinary sexual activity, Tantra teaches

techniques to redirect it upward through the central energy channel (Sushumna) towards higher states of consciousness.

- ◊ "The man who is skilled in controlling the seminal fluid can be said to be the master of Yoga." (Shiva Samhita)

- ♦ Unity of Kundalini and Shakti in Tantra:
  - ◊ Tantra emphasizes the unity of Kundalini (spiritual energy) and Shakti (divine feminine energy). Shakti represents the dynamic creative force of the universe. This union of Kundalini and Shakti is seen as essential for spiritual awakening and enlightenment.
  - ◊ "By the control of the mind, through the power of Yoga, raise Kundalini, the Divine Shakti, who is sleeping at the entrance of the Brahma-dvara (the door to the Brahman)." (The Serpent Power: The Secrets of Tantric and Shaktic Yoga by Arthur Avalon)

These quotes and concepts illustrate the role of Kundalini and sexual energy in Tantra, emphasizing their potential for spiritual growth and the attainment of higher states of consciousness. But it's important to recognize that Tantra encompasses far more than just the exploration of sexual energy. Within the rich fabric of spiritual traditions, Tantra represents a multifaceted path toward spiritual growth, liberation, and the realization of divine consciousness. It offers a holistic approach that integrates the physical, mental, emotional, and spiritual aspects of existence, inviting practitioners to embark on a profound journey of self-discovery and transcendence.

Through its diverse range of esoteric practices, philosophies, and rituals, Tantra serves as a beacon of wisdom and illumination, guiding seekers toward the ultimate realization of their true nature and the unity of all existence.

As we conclude this exploration of the specialized paths of Brahman Vidya and Tantra within Vedanta, we are left with a profound appreciation for the depth and diversity of spiritual practices available to seekers. Now, as we transition to the next section, we will explore the practical application of Vedanta in today's world, learning how its timeless wisdom can guide us in navigating the complexities of modern life and finding fulfillment in our daily existence.

Paths and purpose

As a devoted spiritual student and teacher, I had spent years guiding others on their journeys of self-discovery and spiritual awakening. Yet, little did I know that the most profound awakening of all awaited me during one of my own workshops.

Before I finally decided to include this story, I debated with myself quite a bit about if I was ready to share it publicly and if I was prepared for the judgment that may follow. I am still not sure, but here is the story anyway.

Early in my journey, I had studied with a teacher who had been educated in Tibet. He had initiated me in the study of Tantric texts. When my foundational education was over, he suggested I start leading retreats. Following his guidance, I designed and led my own retreats, mostly focused on purification and breathwork that lies at the heart of Tantra rituals.

Eventually, he suggested I travel overseas to lead a series of retreats one of his past students was organizing and he did not want to travel to. I was quite surprised by this request and also a little nervous, due to the nature of the subject and my relative inexperience. However, as a truly devotional student, I could not say no and accepted to teach this series of retreats.

Fast forward to three months later, I arrived at the location of the last retreat in the series, about an hour's drive from Rio. The workshop was held in a serene location nestled amidst the tranquil beauty of nature — a perfect setting for delving deep into the mysteries of Tantra. As the workshop progressed, I found myself surprisingly drawn to one participant in particular — a woman whose eyes sparkled with a mixture of curiosity and spiritual intensity.

After the workshop ended, she approached me and asked for a 1-1 session the next day, knowing I was staying an extra day. I was very surprised and I have to admit that I felt very conflicted about it. I would need to cross a hard boundary. Up until then, I had followed, and continue to follow, a strictly *no-contact policy* with all my students. However, something intuitively told me it was meant to be. During our private session, I felt a profound spiritual connection developing between us — a connection that transcended the physical and touched the depths of our souls.

Just as we started the initial invocation and breathwork of the practice, I found myself slipping into a trance-like state — a state of pure consciousness where time seemed to stand still, and the boundaries of the physical world dissolved into the infinite expanse of the universe.

For what felt like an eternity, I journeyed to a realm of existence beyond the confines of immediate time and space. In the depths of this elevated state, I found myself transported to an ancient school of mysteries, where I joined the class studying the wisdom of cyclic cosmology.

As I immersed myself in the teachings of the ancients, I felt a profound sense of clarity and understanding wash over me — a deep knowing that transcended the limitations of my physical form. For forty-five minutes (she later told me), I remained in this trance-like state, absorbing the ancient knowledge that was being bestowed upon me.

When I finally emerged from my experience, I found myself filled with a sense of awe and wonder at the profound insights I had received. It was as though I had been granted access to the very fabric of the universe itself, and been given a glimpse into the infinite mysteries that lay beyond.

Now fully back to normal consciousness, as I made eye contact with this woman before me, I knew that our meeting had been no mere coincidence but rather a sacred encounter orchestrated by the divine forces of the universe. And as we parted ways, I carried with me the profound knowing that our connection had opened the door to a deeper understanding of myself and the universe that surrounded me.

I have never met that woman again. I have never initiated anyone 1-1 in these practices again. My *no-contact policy* has and will remain in force. Since then, I have not led a Tantra retreat, either.

And yet, I know I would not be the same but for that experience.

As my mentor later told me, "*When the student is ready, the teacher appears. Often in forms, the student would not expect.*"

Forever grateful to my teachers, I am.

"Vedanta is a philosophy that teaches the unity of all life and the interconnectedness of everything in the universe. This profound insight can help us address many of the challenges we face today, from personal conflict to global issues."

– Michael Nagler

PART V:

LIVING VEDANTA IN TODAY'S WORLD

Vedanta may be an ancient spiritual path but its timeless wisdom continues to serve as a guiding light today for all who choose to follow it. In this final section, we delve into the practical application of the Vedantic principles in the modern world.

As we navigate the complexities of contemporary life, the teachings of Vedanta offer profound insights and practical guidance on how to live a fulfilling and purposeful existence. Drawing from ancient scriptures and the wisdom of enlightened sages, we will explore how Vedanta can empower individuals to find meaning, cultivate inner peace, and navigate today's challenges with wisdom and grace.

From relationships and career pursuits to personal growth and societal engagement, living Vedanta in today's world provides a roadmap for integrating timeless spiritual truths into our everyday lives, fostering harmony, compassion, and spiritual fulfillment in the modern age.

Join me as I share some of the highlights of my personal journey. Although still seeking my own inner truth I hope you will be inspired to unveil the relevance and transformative power of Vedanta in the context of your own life.

"The Vedanta taught me to see that the essence of all religions is the same, that God is within us and that the essence of our being is divine."

– Ralph Waldo Emerson

CHAPTER 17

EMBRACING VEDANTA'S LIGHT

You may have thought that growing up in India meant I had unbridled access to all the wisdom that Vedanta holds. In theory, that's kind of true, although I can't say I was too aware of it at the time. Growing up with a love of knowledge imprinted in my DNA (my father was a History Professor and my mother a Professor of Political Science), I was always encouraged to learn about everything and the academic world was my playground. As for living in a culture that is a melting pot of diversity and tradition, my younger years were infused with ancient rituals and beliefs that I took for granted on a daily basis.

It wasn't until I began studying topics such as quantum physics, psychology (existential), philosophy (determinism), cosmology/mathematics (Brahmagupta vs Pythagoras), and medicine (ayurveda – Swiss Biological medicine) at a later age that I ran into the Vedas again. With my natural curiosity leading me to investigate all aspects of human existence, be that scientific, philosophical, spiritual, or otherwise, it dawned on me that the Vedas were worth revisiting.

To my mind, all of the current knowledge I was soaking up had echoes of ideas and concepts I had previously read about in the Vedas. These sacred texts, estimated to have been composed 1500 BCE (although almost certainly much more ancient in origin), still had a great deal to offer even thousands of years later. That prompted me to reexamine them with greater rigor to understand how this was so.

Having embarked on a lifelong journey to find answers to some of the most profound questions concerning our existence, my spiritual voyage has taken me to some incredible places. I've explored the depths of my soul, tread the great halls of science, immersed myself in spiritual experiences, and traversed the globe. I have humbly sat with acclaimed scholars, brilliant scientists, spiritual teachers, and wizened sages, all of whom have enriched my understanding of the mysteries of life itself.

I have come to believe that what we deem to know is only a grain of sand on an infinite beach; a drop of water in an endless ocean. While we know so very little, Vedanta allows us to glimpse so much more. It's like shining a torch through a crack in the wall of a dark cave to glimpse a breathtaking vista of brilliant light. Its wisdom may seem to be out of reach, far too inaccessible for us to even contemplate accessing it, but all we need is to be curious. It's human nature to want to explore distant horizons, discover new realms, and search for answers to big questions. Vedanta allows us to embark on that quest and even if we never fully grasp the enigma of life's fundamental questions, the journey itself is well worth taking.

Bridging spirituality, new age beliefs, and occult mysticism

Vedanta's connection to spirituality lies in its core teachings regarding the nature of existence and the ultimate reality. Through Vedantic philosophy, we can explore the concept of Brahman, the supreme cosmic consciousness, and Atman, the individual self, recognizing their inherent oneness. These concepts continue to resonate deeply with spiritual seekers who seek to transcend the confines of ego and connect with a higher truth or divine presence. Anyone wishing to connect with something greater than the materialistic and individualistic world we live in can find a deeper sense of purpose and interconnectedness in Vedanta. It provides a framework for understanding that we are all interconnected manifestations of the same divine essence, fostering a sense of unity and oneness among seekers.

Similarly, the exploration of Atman, the individual self, strikes a chord with those on a spiritual quest for self-discovery and inner transformation. In a world where many feel disconnected from their true selves amidst the noise and distractions of modern life, the teachings of Vedanta offer solace and guidance. By recognizing the inherent divinity within themselves and all beings, spiritual seekers are inspired to embark on a journey of self-realization and authenticity, seeking to align their actions and intentions with their higher purpose.

In the realm of New Age beliefs, Vedanta offers a comprehensive framework for understanding the interconnectedness of all things and the power of intention and manifestation. Ideas such as the law of attraction and the power of positive thinking find resonance in Vedantic principles such as the law of karma, which asserts that every action has a corresponding consequence, shaping one's future experiences. This alignment highlights Vedanta's relevance to those seeking to harness spiritual principles for personal growth and transformation.

Regarding occult mysticism, Vedanta provides insights into the unseen realms and the deeper layers of reality beyond the physical plane. Concepts such as Maya, the illusory nature of the material world, and Lila, the divine play of existence, offer parallels to occult teachings on the hidden forces and energies that shape the universe. Vedanta's exploration of the relationship between the manifest and the unmanifest, the known and the unknown, resonates with occult practitioners seeking to penetrate the mysteries of existence and tap into hidden potentials.

If, as I once did (and often still do), you feel that the concepts and ideas within Vedanta are a little overwhelming to grasp, have no fear. This feeling of not knowing enough, or of knowing nothing at all, is quite natural. After all, holding the whole cosmos in your hand and trying to decipher it is no easy task. It takes time, diligence, and patience to glean the wisdom hidden within the details.

This "bankruptcy of knowing" is due to several reasons. Firstly, Vedanta introduces profound philosophical concepts that challenge conventional beliefs and perceptions about reality and the self. This can be disorienting for many, as they grapple with ideas that may seem abstract or paradoxical at first.

Secondly, Vedanta emphasizes the importance of direct experiential knowledge (jñana), rather than mere intellectual understanding. This shift in focus from conceptual learning to personal realization can be daunting for anyone accustomed to relying solely on rational thought and empirical evidence to navigate the world.

Furthermore, Vedanta often reveals the limitations of language and thought in capturing the ultimate truth or reality (Brahman). As we encounter the ineffable nature of existence, we may feel a sense of inadequacy in expressing or comprehending the depth of their experiences.

Additionally, the journey into Vedantic philosophy requires humility and openness to exploring one's deepest beliefs and assumptions about life, which can be unsettling. Confronting the ego and relinquishing attachments to one's identity and worldview are integral aspects of the Vedantic path, but they can also evoke feelings of vulnerability and uncertainty in the early stages of exploration.

All of this is a natural part of the transformative journey toward self-realization and spiritual awakening when encountering Vedanta. It's akin to introducing a four-year-old to rudimentary particle physics, and I don't know many four-year-olds who would be able to handle that without throwing a few tantrums along the way!

But if we immerse ourselves in the teachings and practices of Vedanta with sincerity and patience, they gradually unfold deeper insights and wisdom that illuminate the path ahead.

The gratitude from recognizing the richness of Vedanta's insights and having the opportunity to explore timeless wisdom that offers guidance and solace can be life-changing.

Moreover, sincere engagement with Vedantic practices can lead to realizations that profoundly shift our perspective on existence and the self. Through contemplation and introspection, we may come to realize the interconnectedness of all beings and the inherent divinity that pervades every aspect of creation. This realization fosters a sense of awe and reverence for the magnificence of existence, evoking a deep-seated appreciation of being part of this cosmic dance.

Immersion in Vedantic teachings can awaken us to the timeless truths that lie beyond the fluctuations of mundane, worldly experiences. As we deepen our understanding of concepts such as Brahman, Atman, and Maya, we may experience moments of profound insight and clarity that transcend intellectual understanding. These realizations bring a deep sense of peace and contentment, rooted in the understanding that our true nature is untouched by the transient ups and downs of life.

Through sincere practice and patience, we can all open ourselves to the profound blessings and insights that Vedanta offers, enriching our lives with meaning, purpose, and profound inner peace.

Timeless teachings for the 21st century

We may be living in a fast-paced world of scientific innovation and super-charged technology, but we are still asking the same questions our ancestors did thousands of years ago, albeit in a different context:

- How can I find fulfillment in my work and relationships amidst the demands of everyday life?

- What is the nature of true freedom, and how can I cultivate it in a world filled with obligations and constraints?
- How do I navigate the complexities of morality and ethics in society?
- What role do spirituality and inner growth play in achieving a sense of balance and harmony in my life?
- How can I reconcile the pursuit of material success with the longing for deeper spiritual fulfillment?
- In an era of global connectivity, how can I foster genuine connections and community amidst digital distractions and virtual interactions?
- How can I cultivate resilience and adaptability in the face of uncertainty and adversity?
- What is the relationship between individual identity and collective consciousness, and how does it shape our understanding of self and society?
- How can I overcome the sense of disconnection and alienation prevalent in modern urban life?
- What practices and perspectives can help me transcend the limitations of the ego and cultivate a deeper sense of interconnectedness with all beings?

Such questions reflect our enduring human quest for meaning, purpose, and fulfillment, which resonate with the timeless wisdom of Vedanta, as well as other spiritual traditions. By engaging with these inquiries and seeking answers that resonate with our deepest truths, we can navigate the complexities of the modern world with greater clarity, compassion, and wisdom.

You may have sought answers to these fundamental questions within the confines of a specific religious dogma and have been left feeling dissatisfied. While religions play a vital role in the lives of millions of people across the globe, not everyone finds that they align with their own internal compass. Especially today, where they aren't restricted by cultural norms, geographical constraints, or commonly-held belief systems, many individuals are seeking other versions of truths that go beyond borders or backgrounds.

Vedanta is one path of spiritual development and self-development that transcends the confines of religious dogma while embracing the essence of oneness with all sentient beings. One of Vedanta's distinguishing features is its emphasis on personal exploration and realization, allowing one to embark on a journey of self-discovery and spiritual growth tailored to their own experiences and insights.

Unlike traditional religious systems that may prescribe specific beliefs or rituals, Vedanta provides a framework for understanding the ultimate reality (Brahman) that is accessible to all, regardless of religious affiliation or belief in deities. This inclusivity is empowering, as it invites us to explore the depths of our own consciousness and inner wisdom.

Furthermore, Vedanta emphasizes the interconnectedness of all beings, recognizing the inherent divinity within each individual and fostering a sense of unity and compassion for all life forms. This understanding of oneness goes beyond religious boundaries and cultivates a deep sense of empathy and reverence for the sanctity of all sentient beings.

In embracing Vedanta, one can embark on a path of spiritual development that is free from the trappings of religious doctrine, yet rich in depth and meaning. By delving into the teachings of Vedanta and engaging in practices such as meditation, self-inquiry, and service to others, individuals can cultivate a profound sense of inner peace, purpose, and connection with the world around them.

Spirituality without the need for religion

It is possible to be spiritual without being religious. To understand Brahman, you don't need to believe in any deities, creation stories, or miraculous events. Neither do you need to go to a church or temple, worship icons or idols, nor carry out repetitive rituals. When we look at the differences between Vedanta and belief in a religious god, we can clearly see the fundamental differences in approach and perspective.

In many religious traditions, belief in a personal deity or god is central to the faith. Followers believe in the existence of a divine being who is omniscient, omnipotent, and omnipresent. Worship and devotion are directed towards this deity, often through prayers, rituals, and ceremonies conducted in places of worship such as churches, temples, mosques, or synagogues. Religious texts and scriptures serve as sacred revelations of the divine will, containing stories of creation, miracles, and moral teachings. Followers adhere to a set of doctrines and commandments prescribed by the religion, striving to live according to the will of the deity and attain salvation or enlightenment through faith, good deeds, and adherence to religious laws.

Vedanta offers a different perspective on spirituality that transcends belief in a personal deity or god. At its heart lies the concept of Brahman, the ultimate reality or cosmic consciousness that underlies all existence. Unlike belief in a religious god, understanding Brahman does not require belief in specific deities, creation stories, or miraculous events. Instead, Vedanta invites seekers to explore the nature of reality through introspection, meditation, and self-inquiry, seeking to realize the unity of Brahman within themselves and all beings.

Moreover, Vedanta emphasizes the universality of spiritual truths, recognizing that different religious paths may lead to the same ultimate reality. Unlike religious traditions that prescribe specific rituals and worship practices, Vedanta encourages individuals to find their own unique path of spiritual development, free from external dogma or institutionalized beliefs. While practitioners may engage in practices such as meditation, self-reflection, and service to others, these activities are undertaken as a means of realizing the inherent divinity within oneself and fostering a deeper connection with the universe.

Looking deeper into Vedanta, we discover that although it recognizes the concept of Isvara (personal god), worship in Vedanta is not necessarily directed towards a personal deity. Instead, it may involve practices aimed at cultivating inner peace and focusing on realizing the divine essence within oneself and all beings.

Equally, although Vedanta acknowledges the significance of ritualistic practices, it emphasizes the importance of inner transformation over external rituals. Vedantic rituals may include practices such as puja (worship), mantra recitation, or yajna (sacrificial offerings), but these are viewed as aids to spiritual growth rather than ends in themselves.

In addition, while Vedanta does not reject creation stories outright, it approaches them symbolically rather than literally. Creation myths may serve as allegorical or metaphorical narratives conveying deeper spiritual truths rather than historical or scientific accounts. Sacred texts are acknowledged for their importance in Vedanta, just as in many religions, but they are considered more as sources of spiritual wisdom than doctrines or dogmas. More important in Vedanta is critical inquiry and direct experiential realization over blind adherence to scripture, emphasizing the need for personal understanding and insight.

In essence, while there are overlaps and similarities between religious belief and Vedanta, Vedanta offers a more flexible and inclusive framework for spiritual exploration, focusing on inner transformation, direct realization, and the realization of oneness with the cosmos.

The fountainhead of philosophy

Long before the Greeks were developing complex philosophical works that went on to influence the Western philosophical tradition, Vedic thought was already established in the East. When Greek thinkers such as Thales, Pythagoras, and Heraclitus were deliberating on moral, ethical, and existential questions around the 6th and 5th centuries BCE, the Vedas had already predated them by almost a millennium. That's quite a gap!

As a side note, it's unlikely that the ancient Greeks had direct exposure to the Vedic teachings during their lifetimes. The geographical and cultural distance between ancient Greece and the Indian subcontinent would have made direct transmission of philosophical ideas highly improbable. However, it's worth noting that there were some indirect avenues through which philosophical ideas could have traveled between these regions.

Trade routes connecting the Mediterranean world with the Indian subcontinent could have facilitated some degree of cultural exchange, although the extent to which philosophical ideas were transmitted remains speculative. Following the conquests of Alexander the Great in the 4th century BCE, Hellenistic culture spread across the regions he conquered, including parts of the Indian subcontinent. This period saw increased cultural interaction between Greek and Indian societies. However, the influence of Indian philosophical ideas on Greek thought during this time is not well-documented.

It's possible that certain philosophical concepts or themes found in the Vedas have parallels in other ancient traditions that Greek philosophers may have encountered indirectly. For example, ideas related to the nature of reality, the self, and ethical principles are recurring themes in various philosophical traditions around the world although it's difficult to say exactly how such ideas may have been disseminated at the time.

Nonetheless, Vedic thought holds a significant place in the history of philosophical development and can be considered a fountainhead for several reasons:

Concept of Self (Atman): The Vedas introduced the concept of Atman, the individual self or soul, which is fundamental to understanding human identity and consciousness. This concept forms the basis for much of Eastern philosophy and spirituality and has parallels in Western philosophical ideas about the self and consciousness.

Karma and Ethics: Vedanta offers ethical guidelines rooted in the principles of dharma (duty) and karma (action), providing a framework for understanding human behavior and moral responsibility. These ethical considerations laid the groundwork for ethical systems in both Eastern and Western philosophies, influencing notions of moral responsibility, justice, and the nature of good and evil.

Metaphysical Inquiry: Vedic thought delves deeply into metaphysical questions about the nature of reality, the cosmos, and the ultimate truth (Brahman). This pursuit of metaphysical understanding parallels the inquiries of ancient Greek philosophers and later Western thinkers who sought to comprehend the fundamental principles underlying existence.

Soteriology and Liberation (Moksha): The Vedas introduced the concept of moksha, or liberation from the cycle of birth and death (samsara), through spiritual realization and union with the divine. This idea of transcending worldly suffering and achieving spiritual liberation is echoed in various forms throughout Western philosophical and religious traditions.

Yoga and Meditation: Vedic thought gave rise to the practices of yoga and meditation, which are central to many Eastern philosophical and spiritual traditions. These practices involve introspection, self-awareness, and the cultivation of inner peace and wisdom, serving as foundational elements in the development of psychology and contemplative philosophy.

Cosmology and Nature: Vedic texts contain rich cosmological descriptions and hymns honoring the natural world. The reverence for nature and the recognition of humanity's interconnectedness with the cosmos are themes that resonate with modern ecological and environmental philosophies.

Epistemological Rigor: Vedanta employs sophisticated epistemological methods, including logic (nyaya) and direct experience (anubhava), to explore the nature of knowledge and understanding. This rigorous approach has influenced subsequent philosophical traditions in India and beyond.

Vedanta can also be seen as a bridge between spirituality and modern philosophy, seamlessly integrating spirituality into its philosophical discourse. It recognizes the interconnectedness of the material and spiritual realms and emphasizes the importance of spiritual realization (moksha) as the ultimate goal of human life. This integration of spirituality distinguishes Vedanta from purely secular philosophical systems. Some examples of this are outlined below:

Transcendence of Dualism: Vedanta's emphasis on the underlying unity of existence transcends dualistic divisions between mind and body, subject and object, and material and spiritual realms. This holistic perspective aligns with modern philosophical movements that seek to overcome Cartesian dualism and explore the interconnectedness of all phenomena.

Confluence of Eastern and Western Thought: Vedanta's emphasis on introspection, meditation, and direct spiritual experience resonates with themes found in Western existentialism, phenomenology, and pragmatism. Concepts like self-realization and the nature of consciousness offer insights that complement and enrich modern philosophical discourse.

Dialogue with Science: Vedanta encourages a dialogue between spirituality and science, recognizing that both offer valuable perspectives on the nature of reality. By emphasizing experiential knowledge and inner exploration, Vedanta provides a complementary approach to scientific inquiry, fostering a more holistic understanding of the universe.

Relevance to Modern Existential Concerns: Vedanta addresses timeless existential questions about the nature of human existence, suffering, and the search for meaning. Its teachings on self-awareness, detachment, and the pursuit of spiritual liberation resonate with modern individuals grappling with issues of identity, purpose, and fulfillment.

It's clear that Vedanta serves as a fountainhead of modern philosophy by offering profound insights into metaphysics, epistemology, ethics, and spirituality. Its holistic perspective and integration of spiritual wisdom provide a valuable bridge between traditional wisdom and contemporary philosophical inquiry, enriching our understanding of the human condition and the nature of reality.

From mindfulness to positive psychology

In recent decades, there has been a growing interest in the subject of self-development, with many people searching for greater fulfillment and balance in their lives. Mindfulness practices, therapy, and positive psychology are just some of the ways in which individuals can learn to better handle external pressures and discover inner peace.

With Vedanta offering timeless insights into the nature of the human mind and its potential for growth and transformation, it's only natural that it becomes an attractive proposition for those seeking answers. Its central concepts of self-awareness and introspection are qualities that resonate deeply with contemporary movements such as mindfulness and positive psychology.

For example, Vedanta emphasizes the importance of self-awareness and introspection, which are central to mindfulness practices. Mindfulness involves paying attention to the present moment without judgment, and cultivating awareness of one's thoughts, feelings, and sensations. Vedantic teachings on meditation, contemplation, and self-inquiry provide practical techniques for developing mindfulness and achieving mental clarity and inner peace.

Positive psychology focuses on promoting well-being, resilience, and flourishing by emphasizing strengths, positive emotions, and meaningful engagement in life. Vedanta shares similar goals by guiding individuals toward self-realization, inner fulfillment, and the cultivation of positive qualities such as compassion, gratitude, and equanimity. Vedantic teachings on the nature of the self, the power of positive thinking, and the importance of virtuous living align with the principles of positive psychology.

Vedanta offers insights and techniques for managing stress, anxiety, and negative emotions. Practices such as meditation, breath awareness, and mindfulness help individuals develop emotional regulation skills, reduce reactivity to external stimuli, and cultivate inner resilience. These practices are consistent with mindfulness-based interventions and techniques used in positive psychology to promote emotional well-being and stress reduction.

At the same time, Vedanta emphasizes the importance of inner transformation and self-mastery as essential components of spiritual growth and fulfillment. Similarly, positive psychology recognizes the significance of personal development and self-improvement in fostering happiness and well-being. Both Vedanta and positive psychology advocate for a holistic approach to personal growth that encompasses the mental, emotional, and spiritual dimensions of human experience.

By integrating ancient wisdom with modern science, Vedanta offers a holistic approach to psychological well-being. It elegantly bridges the gap between Eastern spirituality and Western psychology in pursuit of human flourishing by focusing on self-awareness, inner transformation, resilience, and well-being. Through the teachings of Vedanta, individuals can cultivate greater self-understanding, emotional balance, and overall life satisfaction.

The key to living a virtuous life

We all want to live well and do the right thing but often feel conflicted. It's not easy to stay on a virtuous path when the world around us seems to be in a constant state of chaos, exclusion, and overwhelming opinions. How do we stay true to our values and live according to our inner truth amidst the external pressures surrounding us?

Vedanta offers profound guidance on how to live such a virtuous life, placing a strong emphasis on the development of character and the cultivation of virtues as central tenets of its teachings. Here's how it guides us towards virtue development and character purification:

- **Understanding the Nature of Virtue:** Vedanta teaches that virtues such as compassion, honesty, humility, integrity, and selflessness are intrinsic to our true nature as spiritual beings. By recognizing the divine essence within ourselves and others, we are inspired to embody these virtues in our thoughts, words, and actions.
- **Self-Reflection and Self-Inquiry:** Vedanta encourages self-reflection and self-inquiry as essential practices for character purification. By examining our thoughts, motivations, and behavior with honesty and discernment, we become aware of any shortcomings or areas where we can improve. This process of self-awareness

lays the foundation for cultivating virtues and overcoming negative tendencies.

- **Practicing Ethical Conduct (Dharma):** Vedanta emphasizes the importance of following ethical principles (dharma) in all aspects of life. Ethical conduct involves acting in accordance with universal values and moral norms, and treating others with kindness, fairness, and respect. By aligning our actions with dharma, we uphold the principles of virtue and contribute to the welfare of society.
- **Cultivating Positive Qualities:** In Vedanta, virtues are not merely behavioral traits but inner qualities that require cultivation and nurturing. Through practices such as meditation, prayer, and selfless service (seva), we develop positive qualities such as patience, forgiveness, gratitude, and love. These virtues strengthen our character and deepen our spiritual growth.
- **Overcoming Ego and Attachment:** Vedanta identifies egoism and attachment as obstacles to virtue development and spiritual progress. The ego, fueled by desires and self-interest, often leads to selfish actions and moral lapses. By transcending ego-consciousness through practices like self-surrender (prapatti) and selflessness, we can overcome selfish tendencies and live in harmony with others.
- **Seeking Higher Knowledge (Jnana):** Finally, Vedanta teaches that true virtue arises from spiritual wisdom (jnana) and self-realization. By

gaining insight into the ultimate reality (Brahman) and understanding the illusory nature of the ego, we can transcend worldly desires and attachments. This knowledge liberates us from the cycle of suffering (samsara) and inspires us to live a life of virtue and righteousness.

As we can see, Vedanta has a full toolbox that we can use to lead a virtuous life by emphasizing character purification, ethical conduct, and the cultivation of positive qualities. Through self-reflection, ethical practice, and spiritual wisdom, we can overcome egoism, develop noble virtues, and realize our highest potential as human beings.

Embracing Vedanta's Light

In the radiant glow of Vedanta's wisdom, we embark on an extraordinary journey of self-discovery and spiritual transformation, resonating deeply with the challenges and opportunities of the 21st century.

As we tenderly nurture ourselves, we awaken the dormant power of Brahman that resides within, igniting a flame of enlightenment that illuminates the very essence of our being. In a world often fraught with uncertainty and upheaval, Vedanta's timeless teachings offer a guiding light — a beacon of wisdom that transcends cultural divides and speaks directly to the human soul.

With each step, we come to realize the transient nature of all worldly experiences, unveiling the timeless truth that pulsates within us — an eternal source of bliss and liberation. Integrated into the fabric of our lives, Vedanta's teachings become the compass guiding us toward our fullest expression — a life teeming with the vibrant joy of purpose, the serene peace of authenticity, and the boundless freedom of self-realization.

As the Brihadaranyaka Upanishad so succinctly sums up:

"Asato ma sadgamaya

Tamaso ma jyotirgamaya

Mrtyorma amrtam gamaya." (1.3.28)

"Lead me from the unreal to the real

Lead me from darkness to light

Lead me from death to immortality."

This verse beautifully encapsulates the essence of Vedanta's teachings, expressing the aspirational journey from ignorance and illusion to truth, enlightenment, and transcendence.

In embracing Vedanta's luminous embrace, we embark on a thrilling odyssey to discover our most radiant selves, where every moment brims with the promise of divine revelation and profound fulfillment. It enriches our lives and empowers us to navigate the complexities of the modern world with clarity, resilience, and grace.

"Vedanta is not a religion but a philosophy of life, a way of living in the world with an understanding of the oneness of all existence."

– Sarvepalli Radhakrishnan

CONCLUSION: ILLUMINATING THE PATH IN THE ABSENCE OF A GOD

In the mostly contemplative and, hopefully, captivating journey through the pages of this book, we've embarked on an exhilarating quest into the heart of timeless wisdom. From peering beyond the veils of gods and goddesses to unraveling the cosmic mysteries of Brahman and the sacred essence of Atman, we've explored realms of insight, devotion, and profound self-discovery.

In the forest of Vedanta, we've discovered pathways illuminated by the radiant glow of spiritual awakening. Whether traversing the path of wisdom in Jnana Yoga or following the rhythms of devotion through Bhakti Yoga's divine melodies, we've explored the myriad ways to transcendence and inner liberation.

From deciphering karma's intricate dance beyond divine intervention to navigating the eternal cycle of Samsara and Dharma, we've explored the depths of cosmic laws and their profound impact on our journey to enlightenment. As we delved into the arts of Brahma Vidya and Tantra, we've discovered the treasure trove of spiritual knowledge that guides us toward boundless horizons of self-realization.

Now that this enlightening journey within the book has come to an end, my hope is that you will embrace Vedanta's luminous rays and absorb its transformative power in your lives, and the true journey is just beginning.

In the absence of a conventional deity, Vedanta offers a majestic path to self-discovery, liberation, and the unveiling of ultimate truth, which is by far the greatest tool of self realization.

May you continue to tread this brilliant path with fervor, knowing that the pursuit of enlightenment transcends all boundaries, leading you to the possibility of unfolding your greatest divine potential.

My best wishes for the journey forward.

"एकं सद् विप्रा बहुधा वदन्ति"

"Ekam Sat Vipra Bahudha Vadanti."

"Truth is one, the wise call it by many names."
(from the Rig Veda)

ABOUT THE AUTHOR

Dr. Leo Rastogi is a student of life, whose journey seamlessly blends entrepreneurship, philosophy, leadership, research and spirituality. His diverse education, profound spiritual experiences since childhood, and global exploration have shaped his unique world view and insights.

Having traveled to 70+ countries, Leo has delved into ancient traditions, esoteric sciences, quantum physics, and abstract mathematics. He is a distinguished alumni of Harvard Business School, King's College, Oxford, and Salford University, and holds a PhD in Leadership. Leo has also studied psychology, computer science, and neuroscience

As a respected authority on Conscious Leadership, Leo's immersion in monastic practices and his encounters with enlightened masters have given him an especially rich perspective. He bridges the realms of science and spirituality, inspiring scholars, seekers, and students alike. Apart from being an author and much-sought-after public speaker, he is a serial technology and wellness entrepreneur who passionately explores the convergence of emerging tech, science, spirituality, ancient wisdom, and global well-being.

Leo has also served on boards of global business and won numerous awards, as well as contributed to leading publications. He has synthesized much of this wisdom into his critically acclaimed books: "**Many Paths, Many Truths: A Journey through World Religions that have Shaped 20th Century Humanity**" and "**To Discover or**

to Believe: A Journey through what Science does not answer and Spirituality does not ask." He is also the author of "**Ayam: Discover your Beautiful Self**" and "**Life is Good: Conversations about Joy, Purpose, and Fulfillment**".

Leo's life mission is to contribute to transforming human well-being with a synthesis of cutting-edge science, ancient wisdom, and a profound understanding of the human experience.

APPENDIX

Scholarly Texts

The Vedas: The Samhitas of the Rig, Yajur, Sama, and Atharva [single volume, unabridged] by Anonymous (Author), Jon W Fergus (Editor), Ralph T. H. Griffith (Translator), Arthur Berriedale Keith (Translator), CreateSpace Independent Publishing Platform (2017)

The Upanishads by Eknath Easwaran (Author), Nilgiri Press; 2nd edition (2007)

Wisdom of the Ancient Sages: Mundaka Upanishad by Swami Rama (Author), Himalayan Institute Press (2007)

The Philosophy of the Upanishads by Sarvepalli Radhakrishnan (Author), Rabindranath Tagore (Foreword), Edmond Holmes (Introduction), Corner Office Books (2022)

The Complete Book of Yoga: Karma Yoga, Bhakti Yoga, Raja Yoga, Jnana Yoga by Swami Vivekananda (Author), Fingerprint! Publishing (2019)

Tantra of the Great Liberation [Mahanirvana Tantra] translated by Arthur Avalon, John George Woodroffe, Wildside Press (2009)

Hinduism Ritual Reason and Beyond by Ashok Mishra, Story Mirror Publications (2023)

Scientific Research

Creswell, J. D., Lindsay, E. K., & Brown, K. W. (2019). Brief mindfulness meditation training alters psychological and neuroendocrine responses to social evaluative stress. Psychoneuroendocrinology, 106, 138-146.

Keng, S. L., Smoski, M. J., & Robins, C. J. (2011). Effects of mindfulness on psychological health: A review of empirical studies. Clinical Psychology Review, 31(6), 1041-1056.

Jha, A. P., Stanley, E. A., Kiyonaga, A., Wong, L., & Gelfand, L. (2010). Examining the protective effects of mindfulness training on working memory capacity and affective experience. Emotion, 10(1), 54-64.

Tang, Y. Y., Ma, Y., Wang, J., Fan, Y., Feng, S., Lu, Q., ... & Posner, M. I. (2007). Short-term meditation training improves attention and self-regulation. Proceedings of the National Academy of Sciences, 104(43), 17152-17156.

Pascoe, M. C., Thompson, D. R., Jenkins, Z. M., & Ski, C. F. (2017). Mindfulness mediates the physiological markers of stress: Systematic review and meta-analysis. Journal of Psychiatric Research, 95, 156-178.

Luders, E., Kurth, F., Mayer, E. A., Toga, A. W., Narr, K. L., & Gaser, C. (2012). The unique brain anatomy of meditation practitioners: alterations in cortical gyrification. Frontiers in Human Neuroscience, 6, 34.

All images by Veronika By, sourced on Shutterstock.

Made in the USA
Las Vegas, NV
13 October 2025

32252262R00295